C000071752

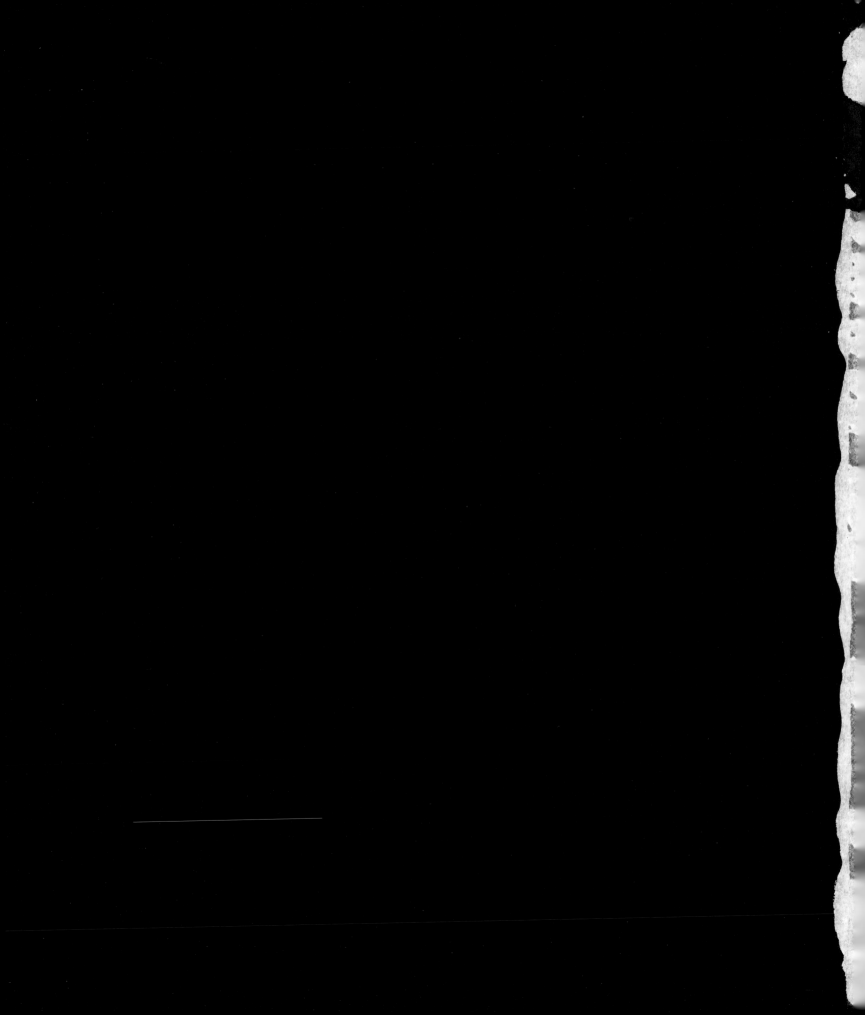

Inspiring
Individuals

Inspiring
Individuals

Ten People Making a Better World

THE ROLEX AWARDS
for ENTERPRISE

Contents

Laureates

Associate Laureates

Preface

PATRICK HEINIGER

Chief Executive Officer of Rolex SA

The spirit of enterprise is 'an essential ingredient in the world's myriad searches for ways to improve the quality of life'. These words were written by my father, the late André J. Heiniger, in the preface to the inaugural Rolex Awards for Enterprise book. Published in 1978, *In the Spirit of Enterprise* introduced the first Laureates of a programme that has become a benchmark in corporate philanthropy worldwide.

Three decades have passed, but the Enterprise Awards are as relevant today as they were in the 1970s in celebrating the creativity, intelligence, determination and courage that define the human spirit. Over the years, the Awards have recognized people who are challenging conventional thinking while seeking to improve the quality of life in their own communities and beyond. These men and women come from all walks of life and from scores of countries around the globe. What they have in common is a great idea and the will to make a difference.

On this occasion of the 13th Rolex Awards, I am more than ever struck by the diversity of spirit presented by the ten Laureates and Associate Laureates featured in these pages. In an age where technological advances have made an indelible mark on society, several have adapted technology to local needs in creative ways: you will learn about a Paraguayan woman who is helping to save forests and resolve her country's housing shortage with an ingenious, eco-friendly solution; a Scottish physicist who is developing a safe, new way to predict volcanic eruptions; an American engineer working to cut harmful emissions from ubiquitous motorized tricycles in Asia; a Jordanian chemist who is conserving ancient Petra; and a Filipino professor whose low-cost invention turns farm waste into affordable domestic energy. You will also read about those safeguarding species and

cultures – from bats in Mexico to the heritage of the tribal communities in remote north-east India; exploring submerged caves in Mexico to gain insight into prehistoric life; and protecting the environment, both India's rainforests and the rich natural heritage of South Africa where a dedicated conservationist is harnessing nature to help Aids orphans find jobs in the ecotourism industry.

These are the fascinating projects that attracted my fellow Selection Committee members, the internationally renowned experts who were inspired by these unconventional individuals. This year's committee was comprised of six new and six 'alumni' jury members, including two who served in the 1990s under André Heiniger. When he created the Rolex Awards in 1976, his intention was to promote the driving force behind our own enterprise – the invincible spirit of mankind. 'By inaugurating the Rolex Awards for Enterprise, we wish to pay tribute and give tangible support to a human characteristic we value,' he asserted.

For more than a century, we at Rolex have been guided by this spirit in all that we do, from our unrivalled achievements in fine watchmaking to the assistance we give explorers, sportsmen and sportswomen, artists and unsung heroes around the world. To echo the sentiments expressed 30 years ago in the first book, it is my sincere hope that the unique stories in this volume will help stimulate new ventures that underscore the essence of what makes us human.

Patrick Heiniger.

RODRIGO MEDELLÍN

Project Location
Mexico

ARTURO GONZÁLEZ

Project Location
Mexico

ANDREW McGONIGLE

Project Location
Italy

TALAL AKASHEH

Project Location
Jordan

MOJI RIBA

Project Location
India

ROMULUS WHITAKER

Project Location
India

TIM BAUER

Project Location
Philippines

ALEXIS BELONIO

Project Location
Philippines

ELSA ZALDÍVAR

Project Location
Paraguay

ANDREW MUIR

Project Location
South Africa

The Spirit of Enterprise

GILBERT M. GROSVENOR

Chairman, National Geographic Society

819 Human endeavour at the highest level, spirit of enterprise, advancing human knowledge, innovation, meaningful discovery – these are all lofty ideas often associated with philosophical phrase-makers, but they are seldom achieved. Fortunately there is an entity – the Swiss watchmaker Rolex – that has thrived on these challenges for more than a century.

More than thirty years ago, to celebrate the 50th anniversary of the extraordinary Oyster chronometer, the late André J. Heiniger, former Rolex chairman, initiated the Rolex Awards for Enterprise to motivate individuals worldwide to improve the quality of life for mankind. Because Rolex has continuously underwritten and nurtured this unique philanthropic programme, it has flourished and succeeded beyond measure.

The Rolex Laureates exemplify the philosophy of the great anthropologist, Margaret Mead, who, when asked if the individual could really make a difference, responded: 'Of course, the individual can make a difference. Indeed that's the only way you can.' The 110 individuals who have received Rolex Awards since 1978 personify Margaret Mead's belief.

Although the Laureates span five different fields of endeavour, they share similar attributes: a passion to help mankind and an exceptional spirit of enterprise. Consider Dr Wijaya Godakumbura of Sri Lanka, a 1998 Laureate in science and medicine chosen for his ingenious safe bottle lamp, which he designed to be spill- and break-proof, thus preventing often-fatal fires in poor homes. Publicity from his Rolex Award has made him known around the world. Hundreds of thousands of these inexpensive lamps have saved countless lives in Sri Lanka and elsewhere in South Asia.

In the area of technology and innovation, Laureate David Schweidenback of the United States was recognized in 2000 for founding Pedals for Progress, which reconditions old bicycles for use in developing countries. When people have 'wheels' they achieve mobility which changes their lives. Job opportunities open up for rural people who have transportation. The prestige of the Rolex Award and public exposure dramatically expanded his market. Today Pedals for Progress has received, reconditioned and donated more than 100,000 bikes in 29 countries, as well as US$10.8 million in spare parts to local partner charities.

The environment category always garners many superb applications. Showing imagination and dogged persistence over many years, Anita Studer's efforts to replant the decimated forests of north-east Brazil rank with the most spectacular achievements. A Swiss ornithologist, Studer originally set out to save a forest that was home to a species of birds she was studying. This 1990 Laureate's project has expanded ever since – not only has her campaign led to the planting of 3 million saplings, it has also produced wide-ranging educational and apprenticeship activities, helping many of Brazil's poorest citizens.

Exploration and discovery attract tenacious visionaries such as Lonnie Dupre, a 2004 Laureate. Seeking to raise awareness about global warming, in 2005 he attempted to cross the Arctic Ocean via the North Pole, but, ironically, melting ice caused by global warming thwarted his efforts. Undaunted, in 2006 he made history by leading a successful summer expedition to the North Pole. To celebrate the 100th anniversary of Robert E. Peary's expedition, Dupre will undertake another epic polar journey in Peary's footsteps in 2009.

The fifth Rolex Awards area of recognition – cultural heritage – supports those who discover, safeguard or contribute to our common historical, cultural or artistic heritage. A 2006 Laureate, Chanda Shroff, in the Indian state of Gujarat, is brilliantly revitalizing the traditional craft of hand embroidery that dates back thousands of years. For 38 years, Shroff, a master craftswoman, has tirelessly brought raw materials, designs and encouragement to villagers. Today, 22,000 women from 120 villages, including all castes, benefit from her visionary concept to resurrect the culture and economy of Kutch in Western Gujarat.

Rolex realized that excellence in the selection of Laureates required the world's best scientists, explorers, educators, ecologists and physicians to judge the 1,500 or so applications that are submitted for each series of the Awards. Consequently, internationally known men and women gather at Rolex headquarters in Geneva to select the Laureates. Judges have included luminaries such as Dr Leo Esaki, Nobel Prize physicist; the late Sir Edmund Hillary; Dr Kathryn

Sullivan, geologist, oceanographer and astronaut; Erling Kagge, polar explorer and mountaineer; heart surgeon Sir Magdi Yacoub; oceanologist Dr Anatoly M. Sagalevitch; Prof. Yves Coppens, prehistorian and palaeoanthropologist; and William F. Reilly, former administrator of the Environmental Protection Agency in the United States. Judging panels combine tremendous knowledge, wisdom and commitment to these unique Awards. Invariably judges continue their relationship with Rolex, urging those people qualified to apply for future awards.

Clearly the driving force behind the Rolex Awards today is Patrick Heiniger, chief executive of Rolex and a visionary himself. He stresses: 'At Rolex we are proud of our long association with human endeavour and excellence at the highest levels, whether in watchmaking, in the arts, sciences or sports….Ingenuity, drive, and quality have underpinned the company since its founding. The Rolex Award winners bear testament to these fundamental principles. I salute them and other inspiring innovators who strive to overcome all obstacles to meet their goals.'

Lastly, Patrick Heiniger, we the judges and all the applicants salute you for your wisdom, innovation, vision and, most particularly, your spirit of enterprise.

Mr Grosvenor was a member of the Rolex Awards Selection Committee in 2000.

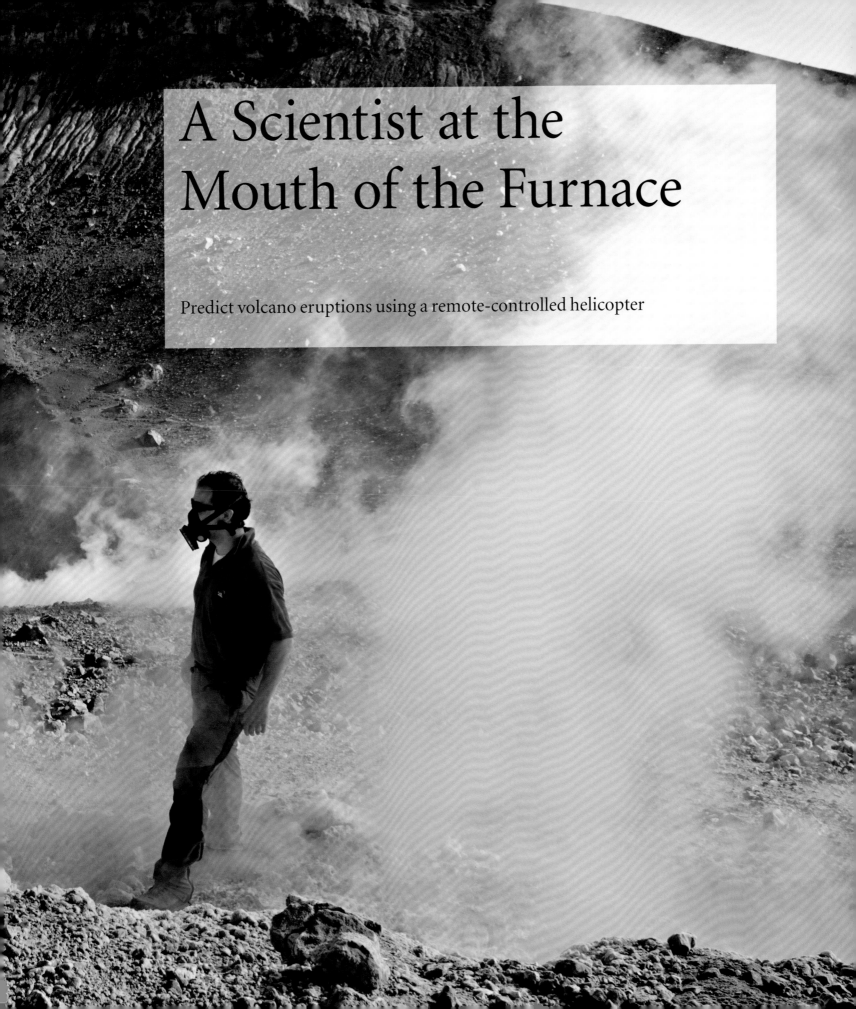

A Scientist at the Mouth of the Furnace

Predict volcano eruptions using a remote-controlled helicopter

Andrew McGonigle UNITED KINGDOM

Many millions of people around the world live in the shadow of an active volcano, at risk of sudden death. Scottish physicist Andrew McGonigle is developing a reliable way to predict eruptions, using an unmanned, small-scale helicopter to measure gases that escape from the volcanic vent. His combination of science and advanced technology has the potential to save thousands of lives.

14 | 15

The ancient Romans named the mouth of Hell *Avernus* – 'birdless' – because of the deadly volcanic exhalations that killed every creature flying over it. Now, a small, man-made bird will approach the vents leading to the underworld, to sample their lethal breath. The knowledge it gathers could help save thousands of lives.

In 2009, Scottish physicist Andrew McGonigle will fly his remote-controlled, 2-metre-long helicopter towards the fiery mouths of Italian volcanoes Etna and Stromboli to inquire of them when they are likely to erupt.

Volcanoes loom large in the human imagination, not only for their vast bulk and fearsome destructive powers, but also for their unpredictability. Their flair for unheralded catastrophe has awed poets, painters, storytellers and scientists alike for generations. Today hundreds of millions of people in many countries dwell in their shadows. Volcanoes present not only a risk to human life, but also a major inconvenience to national and municipal governments – a false alert to evacuate a city can cause unnecessary alarm and much expense; failure to warn citizens in the case of an eruption has far more devastating consequences.

But if Andrew McGonigle's brilliant fusion of high science and smart technology succeeds and is used in conjunction with other measurements specific to each volcano, the fear of sudden volcanic death, which has accompanied humanity since its dawn, will be greatly reduced. Those who live around the 550 volcanoes that have been active at some point in the past may receive weeks, even months of warning of an impending eruption.

To forecast volcanic eruptions, Andrew McGonigle pairs inventive engineering with scientific know-how. Since gas emissions at fuming vents give vital information on possible future activity, he designed miniature sensors and flies them on a remote-controlled helicopter for testing on Sicily's island of Vulcano. A crude sign near the top guides tourists to the crater, which last erupted in 1888.

Researchers have sought ways to predict eruptions for more than a century, a task that sometimes required them to get dangerously close to the volcanoes. While techniques have improved, adequate forewarning remains elusive in many cases. When Mt. Tambora erupted in Indonesia in 1815, it claimed over 70,000 lives. In 1902, Mt. Pelee in Martinique left 31,000 dead and, in 1985, Nevado del Ruiz took 25,000 lives, while Laki killed a quarter of Iceland's population in 1783–84. Yet, in 1991, thanks to measurements of gas emissions, a few days' warning that Mt. Pinatubo in the Philippines was primed to erupt enabled 300,000 people to flee, leaving just 875 casualties. But the gas-based conditions that enabled the forecasting of this eruption are rare: many of the world's volcanoes remain enigmatic and, in the 20th century alone, killed 100,000 people.

One clue to the impending explosion is the release of gases from the magma as it rises towards the surface: sulphur dioxide (SO_2), as well as super-heated steam, and carbon dioxide (CO_2) are the main messengers, bubbling out of the molten rock as the pressure that imprisons them drops. 'It's like popping the cork on a bottle of champagne, allowing the bubbles to escape,' McGonigle explains.

SO_2 is easily detected – but it dissolves in groundwater and is often not released in time to be a reliable indicator of an eruption. The best indicator of all is CO_2, which escapes from the magma much earlier, when it is still 10 kilometres deep in the throat of the volcano. Once it reaches the surface, however, volcanic CO_2 is indistinguishable from atmospheric CO_2 and – until now – measuring methods were not sensitive enough to detect the presence of significant volcanic CO_2. Volcanologists have tried setting their instruments inside the crater, but this method is very hazardous. McGonigle discovered that a better method was to position the instruments below and briefly through the plume of gas as it spewed from the volcano's mouth and floated downwind, high above the ground.

Choosing a remote-controlled helicopter for his project brought McGonigle success. Sampling gases directly is extremely hazardous and experiments with fixed-wing craft by others had met with failure. After about 25 test flights in England, the scientist shipped his aircraft to Italy. Weighing a mere 8 kilograms, the machine is easily carried up the side of the volcano. McGonigle has worked on more than 15 of the earth's 60 active volcanoes.

'Andrew has enthusiasm and drive in abundance and proven ability to deliver a project….he brings many specific skills as well as novel scientific perspectives.'

DR TAMSIN MATHER
OXFORD DEPARTMENT OF EARTH SCIENCES

'Three things are dangerous: tsunamis, earthquakes and volcanic eruptions. This remote-controlled helicopter represents outstanding use of technology.'

DR ANATOLY M. SAGALEVITCH
HEAD OF LABORATORY OF MANNED SUBMERSIBLES, P.P. SHIRSHOV INSTITUTE OF OCEANOLOGY

A simple plastic stool solved an initial problem – failure of the onboard computer. Determining that vibration from the chopper's engine was the culprit, McGonigle suspended his miniature GPS receiver, computer, spectrometer and chemical sensors – a 3-kilogram payload cushioned with foam rubber – on a platform under the seat. He then flew the helicopter into the plume to sample the telltale ratio of sulphur dioxide (SO_2) to carbon dioxide (CO_2).

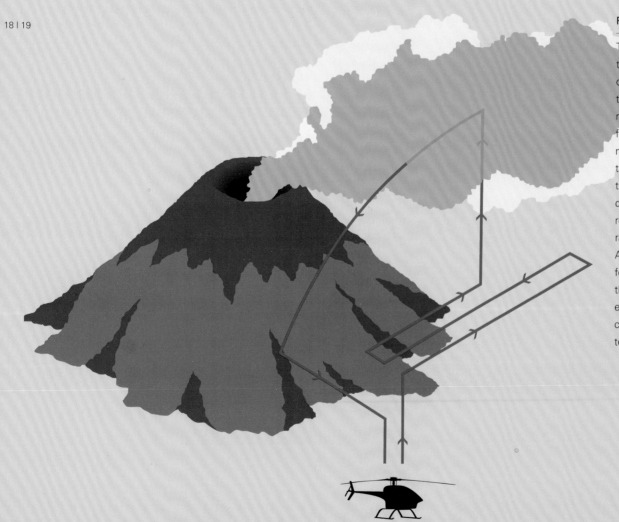

18 | 19

Flight of the Helicopter

The remote-controlled helicopter takes off and makes three crossings below the plume to measure the SO_2 emission rate of the volcano. The craft then flies vertically into the plume to measure the CO_2/SO_2 ratio. Next the helicopter flies briefly through the plume to determine wind direction and speed, which is required for the SO_2 emission rate calculation to be accurate. As the SO_2 flow rate is now known for the previous measurements, the CO_2 emission rate can be easily deduced. Measurements completed, the aircraft returns to the helipad.

McGonigle, who is 35 years old, grew up in Edinburgh, a fountainhead of the earth sciences for two centuries. On cross-country runs and camping expeditions, he skirted extinct volcanoes and explored the rugged, primordial geology of the Scottish Highlands. But his first love was physics: 'I wanted first to understand how the universe worked, through the most fundamental of sciences – although my real passion is to use that understanding to come up with simple, elegant solutions to real problems.'

As a scientist, McGonigle has specialized for a decade in the study of air pollution and volcanic gases using lasers and other sensing devices. Among his adaptations for volcanology is a miniaturized spectrometer, far smaller and less expensive than normal instruments and now in standard use around the world.

He has clambered over 15 of the world's 60 active volcanoes, and analysed the gas signatures of many more. His research has proved seminal, yielding 42 scientific papers that have profoundly influenced volcanology in particular, as well as other areas of research. But it was the coupling of this science with the emerging technology of remote-controlled aircraft that was the stroke of genius, leading to his selection for a Rolex Award.

In 2005, a colleague at Sheffield University, Dr Andy Hodson, began testing a remote-controlled helicopter for studying glaciers, covering a far larger area, more safely and less arduously than on foot. This inspired McGonigle to develop a similar approach for sampling the gases from volcanoes. He rang a model shop, where the obliging manager ran a quick test and told him, yes, a remote-controlled helicopter could carry a payload up to 3 kilograms – enough for the sophisticated sensors needed to take the measurements.

McGonigle teamed up with David Fisher, remote-controlled helicopter champion of Great Britain, and they installed the instruments on a small chopper. To their dismay, the onboard computer that analysed the data failed repeatedly. Eventually they tracked the fault to the massive vibration caused by the helicopter engine – a problem solved with true inventiveness using foam rubber, elastic bands and a US$10 plastic stool. 'For me, the biggest breakthrough was getting it all to work on the test flight. I was almost in tears,' McGonigle recalls.

'I find the whole scale of volcanoes absolutely awesome. They are incredibly powerful phenomena that we cross at our peril.'

ANDREW McGONIGLE

Awkward looking, but effective, AEROVOLC I, as McGonigle named his craft, flies under the sure direction of David Fisher, below at left, Great Britain's remote-controlled helicopter champion. He provided valuable expertise in the selection and testing of the aircraft. Results from Vulcano proved that unmanned helicopters are ideal for volcanic gas measurements, but an easier-to-operate craft with greater range is needed.

In March 2007, with the help of David Fisher and Professor Alessandro Aiuppa, of the Italian National Institute of Geophysics and Volcanology, the prototype helicopter, AEROVOLC I, took to the skies near the fuming vent of Vulcano, a modest cone near Sicily that has lent its name to the entire volcano tribe. It worked perfectly – over the ensuing days the instruments recorded SO_2, CO_2 and wind speed, enabling the scientists to calculate the flow of gases from the volcano. 'It was just amazing. There was ecstasy in the camp. We had clear proof that the concept worked – and that it may be possible to predict from weeks to months ahead whether an eruption is developing, from the flow of CO_2,' says McGonigle. Professor Aiuppa adds: 'These measurements could provide us with the earliest and most direct possible indicator of a forthcoming eruption. McGonigle's idea is innovative and should represent a major break-through in modern volcanology.'

The method requires the team to measure the flow of SO_2 from beneath, then fly the helicopter into the plume to measure total SO_2 and CO_2 and establish the wind speed, all with the aim of accurately calculating the flow rate of volcanic CO_2. This provides clues to the state and position of the magma, deep in the volcano – advance notice that 'something is going on'. However, McGonigle cautions, each volcano is different – special knowledge of its unique 'personality' must be added to the information about its gas emissions.

'These measurements could provide us with the earliest and most direct possible indicator of a forthcoming eruption. McGonigle's idea is a breakthrough.'

ALESSANDRO AIUPPA
GAS GEOCHEMIST AND VOLCANOLOGIST, UNIVERSITY OF PALERMO AND INGV

With the Rolex Award funds, McGonigle is purchasing a piece of high technology, a 14-kilo helicopter built by an American firm, which will be known as AEROVOLC II once he has equipped it with his specially tailored gas sensors and analytical software. In 2009, he will carry out further field trials on two of Italy's most famous but very different volcanoes, Mt. Stromboli and Mt. Etna: the former erupts every 10 minutes and the latter about once a year. Using GPS navigation and onboard robotics, the helicopter can take off, fly and land itself according to a pre-determined flight plan, enabling anyone with basic technical skills to operate it. Or, thanks to an onboard video camera, it can be guided manually at any stage of a flight extending up to around 20 kilometres. This makes the technology usable with minimal training by the staff at any volcano observatory in the world. Measurements can be taken safely, cheaply and frequently, even daily, replacing, for example, current piloted helicopter flights over volcanoes that are expensive and at times perilous.

'People are now interested in the prospect of using these kinds of aircraft for all sorts of monitoring and mapping, for anything that is remotely dangerous,' McGonigle adds. The equipment costs about $80,000, a fraction of the price of other, less versatile approaches. Thus, for relatively modest costs, millions of people could receive the gift of time in which to save themselves from an impending eruption.

Combined with other forms of volcanic sensing, such as seismology and ground-deformation detection, Andrew McGonigle's innovative approach to gas sampling will provide far greater power and precision in predicting the unpredictable, the timing of a volcanic outbreak – and so protect countless people from an age-old, terrifying menace to humanity. |

Demonstrating his user-friendly software, McGonigle shares his test results, top right, with Italian volcanologists Dr Sergio Gurrieri and Prof. Alessandro Aiuppa. With their support and his Rolex Award, he plans to test AEROVOLC II, with a range of 20 kilometres and double the payload – including a new sensor he shows Aiuppa and electrical engineer Gaetano Giudice, below. In 2009, McGonigle hopes to measure the gases of Stromboli, right, and Mt. Etna, two of Italy's most famous volcanoes. Success will bring protection for hundreds of millions of people who live next door to active volcanoes.

Julian Cribb WRITER   Marc Latzel PHOTOGRAPHER

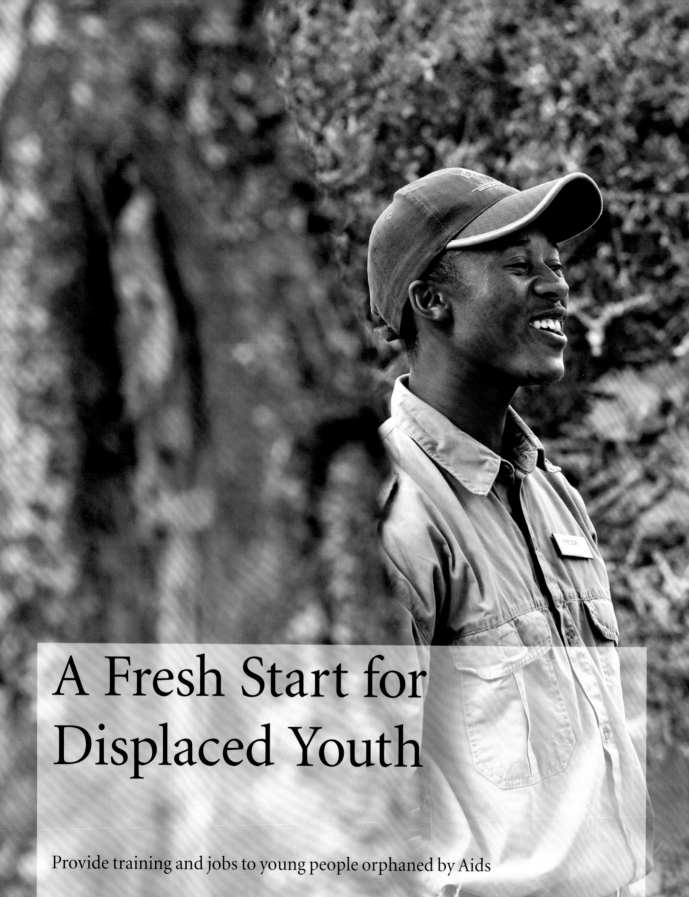

A Fresh Start for Displaced Youth

Provide training and jobs to young people orphaned by Aids

Andrew Muir SOUTH AFRICA

South African conservationist Andrew Muir is harnessing the healing powers of nature to help young people orphaned by HIV/Aids become independent citizens. Muir's Umzi Wethu programme provides vulnerable but motivated youths with vocational training and jobs in the burgeoning ecotourism industry, while immersing them in their country's rich natural heritage.

28 | 29

Pakamisa Kolisi was just 13 years old when his mother died. At the time, he lived in a cardboard shack in the sprawling township of Zwide, just outside Port Elizabeth in South Africa's Eastern Cape Province. His father had abandoned him long before, so the mother's death meant Pakamisa was an orphan. His seven-year-old brother Mandilakhe now looked to Pakamisa for food and clothes, and to get him ready for school. For a while, Pakamisa was helped by his grandmother, but she too became frail and died. His future looked bleak, particularly as he lived in a township where most of the 35,000-plus inhabitants lived below the poverty line, were unemployed and lacked proper housing.

Pakamisa's story, sadly, is not an unusual one. South Africa has an estimated 88,000 child-headed households. As the nation continues to suffer the devastation of an Aids pandemic – South Africa has the biggest number of HIV/Aids infections of any country in the world, with 5.5 million people believed to be infected – the number of children with no parents is expected to increase significantly over the next decade. The death of a parent by HIV/Aids accounts for about half the country's estimated 2.2 million orphans. Traditional family structures are becoming overstretched, unable to cope with the overwhelming number of children with no adults to care for them.

Getting a taste of their future, young South Africans cook, serve and wash up at Conyngham's Coffee Shop in Port Elizabeth, operated by Umzi Wethu, a programme conceived by Andrew Muir to rescue young people impacted by HIV/Aids and other orphans. Saved from a life of poverty, orphan Pakamisa Kolisi, right, trains to be a junior chef.

'Umzi Wethu is absolutely original. There is nothing currently like it in Africa. The model … needs to be expanded and repeated.'

NANCY GELMAN
AFRICA BIODIVERSITY COLLABORATIVE GROUP

In most developing countries, orphaned children are highly vulnerable. Not only have they lost the security and moral guidance that parents provide, they are also at risk, in poorer countries, of being denied shelter, food, clothing and health care. Many are unable to attend school as they have to look after younger siblings or contribute financially to the household. Understandably, such children often lapse into depression, develop a dependence on alcohol or drugs, or turn to crime or prostitution to survive. To escape such a cycle of poverty is virtually impossible.

What is remarkable about Pakamisa's story is that, today, at the age of 25, he is content. Thanks to the Umzi Wethu Training Academy for Displaced Youth, he has embarked on a career in the hospitality industry and is well on his way to becoming independent.

The brainchild of South African conservationist Andrew Muir, Umzi Wethu is a multifaceted intervention programme that targets orphans. Aged 42, Muir vividly recalls the day the seed was planted for Umzi Wethu: 'I read a UN report back in 2004 that 80 per cent of the world's orphans live in sub-Saharan Africa. I was shocked. This is a massive issue, not only from a social perspective, but also from an environmental perspective. I understand the pressure this can have on the environment, particularly in very poor regions. Orphans are vulnerable, and generally have no other option than to use the resources readily available to them. This can lead to poaching, chopping down of trees for firewood and shelter, and the like.

'I also realized that conservationists as a collective have not come up with a solution, one that could make this crisis take a positive turn. Conservationists today need to be aware of the broader social context within which they operate. Gone are the days when we could simply put a fence around a protected area and ignore what was happening outside.'

As executive director of the Wilderness Foundation South Africa (WFSA), a non-profit conservation organization that uses nature as a tool for social change, Muir is aware of the healing capacity of nature and the significant employment opportunities offered by ecotourism. Over the course of his career, he has created dozens of initiatives and events linking conservation causes with social development, and has raised over US$26 million for conservation and social development programmes. Umzi Wethu, which means 'Our Home' in Xhosa, one of South Africa's 11 official languages, is Muir's latest project, a holistic programme that aims to fulfil the potential of resilient, motivated youths like Pakamisa who have been displaced by HIV/Aids and poverty.

Practising his kitchen skills, Pakamisa learns how to prepare butternut squash, above, a favourite vegetable in his country. Umzi Wethu will also provide mentoring, wellness counselling and wilderness experience – Andrew Muir, who talks to Pakamisa at far left, believes in the healing power of nature. Pakamisa visits Pinky Kondlo, project manager for Umzi Wethu, left, and fellow student Ntomboxolo Ngoxoza.

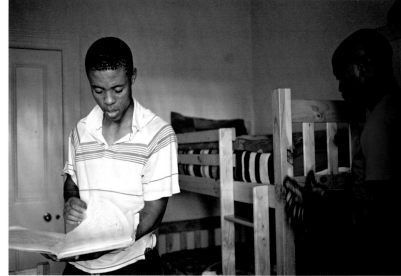

Muir primarily targets orphans from the most impoverished communities, providing a secure, nurturing place for day and residential students to develop. About 60 per cent of the youths attending Umzi Wethu are 'Aids orphans' – individuals impacted by Aids but who are not necessarily infected by the virus – while the remaining intake comprises individuals living in child-headed households. Muir stresses that Umzi Wethu is neither an orphanage nor a state institution, but a training facility that teaches teenage orphans to be cooks or rangers, the two sectors in the eco-tourism industry with the highest potential for income and management opportunities.

'We don't choose individuals directly, but work with institutions who identify potential candidates and who can provide us with a case history. Over and above vocational tuition, Umzi Wethu also provides life skills training, wellness counselling, one-to-one mentoring and wilderness experience. I knew that the only way we could make Umzi Wethu work was by creating a comprehensive and long-lasting programme. After all, the vulnerability of these orphans has generally been 18 years in the making and will need something pretty intense and all-embracing to turn it around!'

As well as a year of training, Umzi Wethu provides guaranteed job placement with transitional support. As the costs of any programme like Umzi Wethu are substantial, Muir believes it is essential to select young people who have the aptitude and potential to see the programme through. 'We need to be confident that our Umzi Wethu students will be able to stay in the jobs we secure for them – only then will the programme be successful from a socio-economic perspective. Essentially we are creating ambassadors for other vulnerable youths and orphans to look up to, and see a brighter outlook. Graduates from Umzi Wethu will serve as examples of opportunity and hope to both their own families and to the communities from which they come.'

Living and learning together lead to success for student residents, many of whom enjoy a safe home for the first time, thanks to Umzi Wethu, which means 'Our Home'. Not all study to be chefs. Seven graduates, celebrating with Muir, top left, will become rangers in wildlife-rich South Africa. 'The 29 parks and 20 private game reserves in the Eastern Cape generate jobs that often pay twice the minimum wage, but demand skilled labour,' says Muir.

'There is increasing evidence that nature has a beneficial and therapeutic effect on the human psyche. It is brilliant to combine securing work for orphans with conservation goals.'

DR GEH MIN
OPHTHALMOLOGIST, ENVIRONMENTALIST AND PRESIDENT OF THE NATURE SOCIETY OF SINGAPORE

'Nature, nurture, future,' is Umzi Wethu's motto, a promise that has changed the lives of two ranger graduates thanking Muir, below left. They work in Shamwari Game Reserve where visitors can see the big five: elephant, rhino, buffalo, lion and leopard.

There is plenty of room for animals to roam at the new 15,800-hectare Kuzuko Game Reserve (below). Offering luxury lodges and bush camp, the reserve will employ both chefs and rangers.

South Africa's Eastern Cape Province, where Muir piloted the Umzi Wethu programme, boasts remarkable natural diversity. Ironically, far more Western tourists have had more access to these wilderness areas than the people who live nearby. Muir believes that 'the wilderness can heal and sustain the human psyche'. All Umzi Wethu students spend approximately one week every two months off-site at bush camp to experience, often for the first time, South Africa's great wildlife and plant biodiversity. For a young orphan, used only to the muddy, littered streets of a shanty town, to walk through the bush or next to a running river in a forest can be life-changing. And for this same young person to be surrounded by peers from a similar background, in a setting that is secure, with boundaries of discipline set by a caring House Mother, will bring major psychological benefits. Add credible vocational training and you have the ingredients that make up Umzi Wethu, transforming such young people into effective, independent citizens. This pioneering method of using conservation to rescue orphans and vulnerable youths from poverty and despair has won Andrew Muir a Rolex Award.

Initial studies and assessments by the Wilderness Foundation indicate that in South Africa alone more than 50,000 new jobs are being created every year in the conservation and tourism industries leading up to a major football event – the 2010 FIFA World Cup, which South Africa will host. The Eastern Cape is poised to reap the benefits of this upsurge in ecotourism, and two Umzi Wethu academies have been established there. The first, in Port Elizabeth, focuses on hospitality, while the second, located in Somerset East, close to three national parks, caters exclusively to the conservation and ranger trainees. Funds from Muir's Rolex Award will help cover the costs associated with one intake of 20 students at the Somerset East academy. The success of the pilot programme at the two academies has been outstanding, as Muir proudly explains: 'Eighty-five per cent of the 40 graduates have made the successful transition into employment, with some already having been promoted to junior managers.'

Umzi Wethu is, for Muir, also a means by which he can help heal the wounds left by apartheid. It was his exposure to the atrocities committed

'I now know where I am going. My future is bright and secure, and I really believe that I can fulfil my dream to be a manager in a hotel.'

PAKAMISA KOLISI
UMZI WETHU STUDENT

by the military in suppressing the anti-apartheid movement that flamed his desire to redress the longstanding racial injustices in South Africa. Conscripted into the South African Defence Force (SADF) in 1985, he was deployed into the townships to help quell the uprisings. 'This marked a turning point in my life,' recalls Muir. 'I witnessed some horrific things, and saw absolute anger and hatred in the eyes of the majority of our country's citizens. I knew then that I had to try, in some way, to heal those wounds.'

In fact Muir's experience in the army showed him not only the worst side of the country, but also its best. Having conscientiously objected to his duties, he was, after a period of incarceration, assigned a desk job. It was here that he conceived the idea of a 780-kilometre walk along the beach from Nature's Valley, near Plettenberg Bay, to Cape Town to raise awareness about tuberculosis. Under the auspices of the SADF, Muir was joined by 13 of the country's premier athletes on a walk that took 28 days. He says the experience transformed him, inspiring him to develop his philosophy of using nature for both social and environmental reform.

His deep commitment has won wide respect, as Dr Mamphele Ramphele, a leading intellectual and political figure, acknowledges: 'Andrew is a man fired by the passion to be the best he can be in order to make the world he lives in a better place. He is a leader, teacher, mentor and bridge between young people and the world of adults; between black and white South Africans; and between urban development and the relatively unspoiled wilderness areas. He is able to move people beyond the limits they set themselves.'

Muir wants to create 10,000 jobs over the next 10 years. To achieve this, the Umzi Wethu model needs to be embraced throughout the Southern African region. 'Umzi Wethu was never created as a one-off programme, but rather as a model that can be duplicated in other provinces and countries, and in other industries. My hope is that the Rolex Award will be the catalyst for rolling out this programme more widely, to benefit the millions of orphans living in Southern Africa.'

Pakamisa says Umzi Wethu has changed his life. With a job as a junior chef, he is better placed to look after his brother Mandilakhe. 'I now know where I am going,' he says. 'My future is bright and secure, and I really believe that I can fulfil my dream to be a manager in a hotel. What I pray for is that my friends living in the township will also be able to share the Umzi experience, and be given the chance to live their dream.' |

New opportunities await three graduate rangers at the Kuzuko Game Reserve. Muir's second academy in nearby Somerset East will be exclusively for training rangers in skills such as animal tracking. Identification of paw tracks at the Shamwari Game Reserve leads to a cheetah in the grass. Muir hopes to duplicate the success of Umzi Wethu, not only elsewhere in South Africa, but also in Mozambique and Namibia.

Alexa Schoof Marketos WRITER   Tomas Bertelsen PHOTOGRAPHER

Building the Loofah House

Combine loofah and plastic waste to make low-cost housing

Elsa Zaldívar PARAGUAY

Elsa Zaldívar takes leftover pieces of a vegetable sponge and mixes them with other vegetable matter and recycled plastic to form strong, lightweight panels that can easily be assembled into simple structures, including houses. Her technological, eco-friendly solution to her nation's housing shortage will help save what remains of Paraguay's rapidly diminishing forests.

40 | 41

In the poverty-stricken countryside of Paraguay, a landlocked country in the heart of South America, an innovative social activist has found a new use for an old vegetable. Elsa Zaldívar, whose longstanding commitment to helping the poor while protecting the environment has won her deep respect in her native land, has found a way to mix loofah – a cucumber-like vegetable that is dried to yield a scratchy sponge for use as abrasive skin scrubber – with other vegetable matter like husks from corn and caranday palm trees, along with recycled plastic, to form strong, lightweight panels. These can be used to create furniture and construct houses, insulating them from temperature and noise. About 300,000 Paraguayan families do not have adequate housing.

Elsa Zaldívar was born in Asunción, the nation's capital, in 1960, during the repressive 35-year rule of President Alfredo Stroessner. Her mother was an artist and her father a committed political leader fiercely opposed to the military dictatorship. Zaldívar inherited their passion for change and became involved in social programmes, working with poor people in her neighbourhood. She took a degree in communications and, from 1992, ran a rural development programme in Caaguazú, a region that had experienced severe deforestation for more than four decades. Her work quickly showed her how making a simple change can transform people's lives.

Strangely, the first step in house building for Elsa Zaldívar is growing a vegetable used for bath sponges. She and Teodora Arguello, head of an association of sponge growers, right, encourage rural women to grow high-quality loofahs. What to do with material left after creating the bath products led Zaldívar to combine the waste with recycled plastic to make panels for inexpensive housing.

'Zaldívar is very practical. This is a very creative, down-to-earth project which deals with a serious social problem in Paraguay.'

YOLANDA KAKABADSE
ENVIRONMENTALIST AND GLOBAL CHAMPION OF SUSTAINABLE DEVELOPMENT

Waste that can be turned into resources is a mine for Zaldívar. In the region of Caaguazú, where she began her project, piles of loofah debris on the floor, left, will be gathered and mixed with plastic from a sorting centre, top, next to the metropolitan landfill on the outskirts of Asunción, Paraguay's capital. Tons of discarded material are deposited here daily, 66 per cent of which is biodegradable.

'We carried out a project with women to construct toilets, and we built stoves for them to cook on,' she explains. 'It was impressive how these simple acts changed the women's lives. They told me: "Now we feel like we're people with dignity." That's the result of simply having a bathroom inside or close to the house rather than 100 metres away, and being able to cook on a stove rather than stooped over a fire on the ground.'

Zaldívar decided that the most effective way to improve the lives of rural women was by increasing their earning capacity. The area's economy had declined with the collapse of cotton and increasing cultivation of soya, an environmentally disastrous crop that had left soils contaminated and forced families off the land, leaving them without employment. Zaldívar took an interest in loofah, a plant that grows easily in the region, but which had fallen out of favour. She persuaded local women in Caaguazú to consider it as a means of generating income.

When harvested before it completely ripens, loofah can be eaten. But Zaldívar's women let the plant ripen and dry out, then process it until only a fibrous sponge remains. Their hard work, along with the ecological methods they used and the quality of the fibre they produced, gave the product a competitive advantage over plantation-grown loofahs from China and other countries. The women organized themselves in a cooperative and sold their loofah sponges as cosmetic products. They used loofah to manufacture mats, slippers, insoles and a variety of other products that were exported to markets as far away as Europe. The women's earnings grew, and the successful enterprise drew praise from environmentalists and others. Eventually it even won the respect of the local men who had initially laughed at the project as a women's idea that had no chance of succeeding.

Zaldívar wrote a manual about growing loofah to spread the word to other regions. She was awarded an Ashoka Fellowship in 2001 to continue her efforts to empower rural women to make products from loofah products and to set up a micro-enterprise.

Yet Zaldívar was not fully satisfied with the cooperative's success. Even with the women's efforts to grow the high-standard vegetables, roughly one-third of the loofahs they cultivated were of inferior quality and could not be exported. And another 30 per cent of the sponge material destined for the finished products was trimmed off during manufacture. Determined to find a market for the loofah waste, Zaldívar teamed up with Pedro Padrós, an industrial engineer, to search for a way to use the vegetable material to construct inexpensive panels for walls and roofing for building houses. She had realized that if the first step to improving the lives of the poor was increasing their income, the next was to help them find decent housing, which would dramatically raise their living standards. Zaldívar was highly enthusiastic. But, disappointingly, initial efforts to mix loofah with different types of glues did not produce the desired result, mainly because of the high costs involved.

Then Padrós got the idea of using plastic waste with the loofah. He invented a machine that melted a mixture of three types of recycled plastic and combined the resulting liquid with loofah and other vegetable fibres, such as cotton netting and chopped corn husks. After hundreds of trials, they succeeded in obtaining a working product. With help from Paraguay's environment ministry, Base ECTA (a non-profit organization headed by Zaldívar) obtained a grant from the Inter-American Development Bank to construct the prototype of a machine to produce the panels.

Zaldívar's vision plus waste plus the invention of a new machine produce a sheet of panelling that is flexible and tough, and can insulate from heat and noise. Loofah fibres and two kinds of plastic, below, will be combined and laminated onto a cotton roll, right, which will eventually be extruded as building panels.

'This project addresses the problem of housing that in many countries causes illness and at the same time it can reduce the exploitation of wood.'

BERTHA CAMACHO
SWISS RESOURCE CENTRE AND CONSULTANCIES FOR DEVELOPMENT (SKAT) IN ST GALLEN

Too many trees have been cut down to provide land for a largely agricultural Paraguay. Before a zero deforestation law was passed in 2004, 90 per cent of the country's virgin timber had been cut down and regrowth is slow. Farm houses, opposite top right, built of wood can now be constructed of the composite panels, as seen in a model house, left. Pedro Padrós, who designed the manufacturing machine, demonstrates the material's flexibility for Zaldívar, above.

Combining a melting unit, mixer, extruder and cutting unit, the machine can produce – in an hour – a half-metre-wide panel 120 metres long. Depending on the exact mix of plastics and fibres, as well as the thickness of the panel, the composite can have varying amounts of flexibility, weight and insulating qualities, making it adaptable to a variety of construction needs. Colouring can be included in the panel's plastic mix at the time of fabrication, so there is no need to paint the walls after construction, saving homeowners time and money. Padrós says a panel of even greater strength can be created by using a honeycomb or earthen filler, as well as vegetable matter, to create a sandwich of two panels.

The composite panels are easier to handle than lumber or brick, and much better than conventional materials in an earthquake or other natural catastrophe. Combined with special metal connectors, 'it will bend but not break', Padrós says. And if a house does collapse, he says, someone is much more likely to survive if the walls are lighter in weight than conventional materials. Using the panels will also help spare the nation's forests. 'Because we're using fibres that are completely renewable, we can stop using lumber for construction. That's very important in Paraguay because we've already reduced our original forest to less than 10 per cent of Paraguay's territory,' Zaldívar points out. 'We're running out of trees.'

As Padrós has refined the design of the panels, improvements have brought the cost down. The panels initially cost about US$6 per square metre to produce, but the cost has already dropped to less than half that figure, making it competitive with existing construction materials, such as wood. Zaldívar predicts the price will continue to fall as experiments continue. She is also involved in discussions with several companies interested in using the panels commercially, but her main aim is to make the material available at low cost to those who need it most.

By supplementing the panels with other locally obtained materials such as bamboo and adobe, Zaldívar believes rural families should be able to build their own simple house in just three to four days. Even urban residents, who often have access to subsidized credit and other government assistance, will be able to use the panels in constructing decent housing.

The project's success derives from the unique combination of Padrós' engineering skills with Zaldívar's genius in creating an integrated system of cultivation, recycling, production and distribution. In addition to the loofah producers, Zaldívar is working with recyclers in urban areas in order to guarantee a flow of appropriate plastic, and with groups of women to provide the tonnes of corn and palm husks, for example, that will be needed – all materials that would otherwise end up in landfills.

Padrós says the panels are designed so that they will not generate any waste – should they wear out or break, they can be ground up and recycled into new panels. The process could be repeated several times until the composite becomes too rich in vegetable fibres, but Padrós says the mixture can then be used as a high-energy fuel. That means the recycled plastics used in the initial mix must be carefully selected to ensure that they can be burned without producing toxic fumes.

Paraguayans are greeting news of the panels with excitement. Gustavo Candia, a Paraguayan consultant on good government and poverty reduction for the German development organization Gesellschaft für Technische Zusammenarbeit (GTZ), says that Zaldívar's initiative allows 'primary producers to participate in adding value to their products', a distinct achievement for poor rural farmers. Zaldívar's reputation as an innovator is well deserved, he believes. 'With this project, Elsa reaffirms that with persistence and reflection, you can create socio-economic impact in sectors that have generally been excluded from progress,' Candia said.

As Zaldívar and Padrós finish testing the improved panel-making machine, the Rolex Award will finance a promotion centre near Asunción and the construction of three model houses where the panels' versatility will be displayed for both urban and rural audiences, as well as funding the production of a video that will be used to describe the project to people interested in using similar techniques in other countries.

Zaldívar's initial focus for providing low-cost housing remains Paraguay's deforested countryside. 'We want to find sustainable housing alternatives for the poor, while also discovering new markets for their agricultural products, particularly the loofah. This is a perfect combination,' she says. |

A woman determined to help the people of her country find a better life, social activist Elsa Zaldívar is a communicator who has become knowledgeable in growing sponges, bamboo construction and gender education. Now she and her associates are producing a new building material that promises a practical remedy for housing for the poor of Paraguay.

Paul Jeffrey WRITER   Jess Hoffman PHOTOGRAPHER

'To have a dignified home liberates people. They realize they can live in another manner. It allows them to move forward in other parts of their lives.'

ELSA ZALDÍVAR

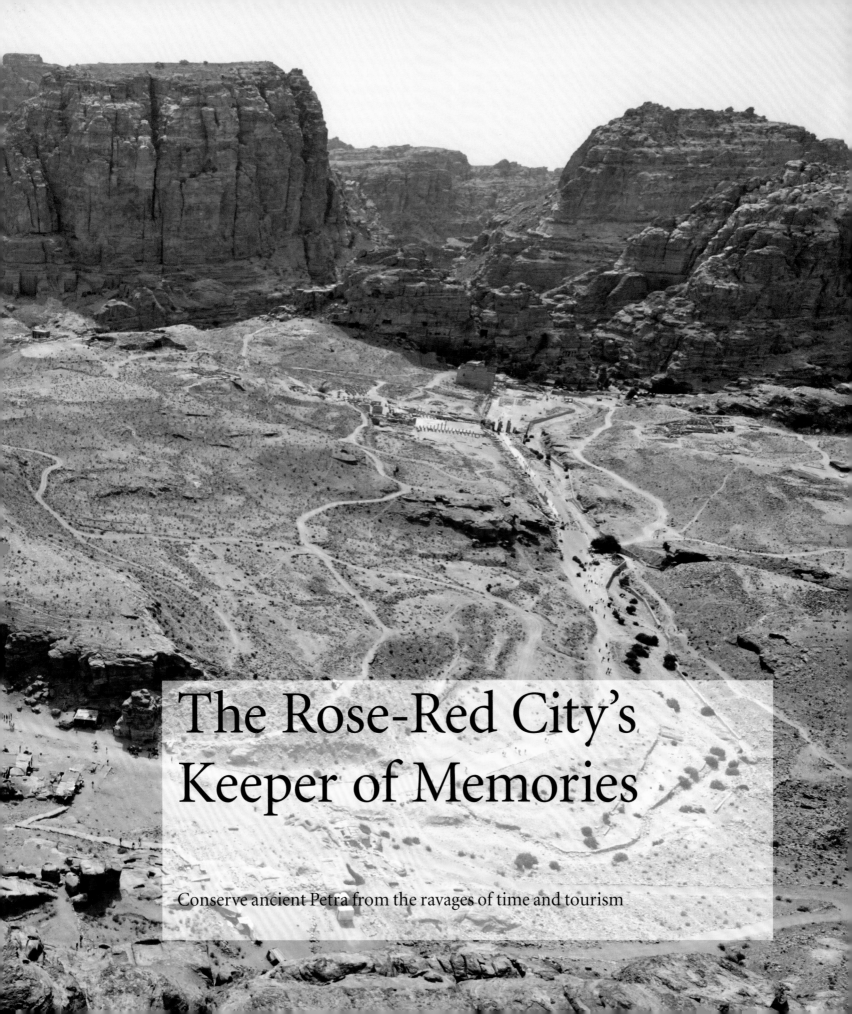

The Rose-Red City's Keeper of Memories

Conserve ancient Petra from the ravages of time and tourism

Talal Akasheh JORDAN

Petra, a 2,500-year-old city in Jordan, is one of the world's most revered cultural sites. But time and tourism are taking their toll on its monuments, carved into sandstone cliffs. Chemistry professor Talal Akasheh, who has devoted 26 years to documenting Petra scientifically, is set to complete a knowledge system that will underpin the city's future conservation.

Like a city carved from a sunrise, Petra rests among the desert sands, recalling the genius of creators long lost in time. Yet the very beauty that adorns it is also, slowly, devouring it.

The chemistry that decks Petra's ancient sandstone monuments in red, sulphur and orange hues is also secretly, grain by grain, dissolving this ancient wonder. Bone-stripping desert winds, rain and flash floods, scorching sun, teeming tourists and modern development add their weight to the erosion. Against these irresistible natural and human forces stand the resolution and skill of Talal Akasheh, a man determined to help save Petra for future generations.

Since he first beheld Petra as a young scientist, the soft-spoken professor of chemistry from Jordan's Hashemite University has dedicated 26 of his 61 years to conserving the city. As he marvelled at the magnificence of its ancient monuments carved from the living rock by a forgotten people, he also noted with distress the ravages of time.

'I was astonished by the beauty of the site, its geology, its architecture. But I also saw many signs of deterioration. I felt something should be done about it,' he explains. 'It is a place alive with history – including the history of my own family, originally from Petra. I could not look at such beauty without saying: "Maybe I can help. Maybe not a lot, but perhaps a little, I can help".' With quiet persistence, scientific meticulousness and inspired leadership, Akasheh has helped piece together the knowledge system that may yet help protect Petra, or at least delay its complete destruction for as long as possible.

Wonder of the ancient world, the rock-cut temples, tombs and amphitheatre of Petra have a protector, Talal Akasheh, a chemistry professor who for 26 years has worked to preserve this remarkable site. Now he has enlisted archaeologists, engineers, physicists and hydrologists to help him produce a comprehensive database that will be a vital resource in safeguarding this extraordinary jewel in the Jordanian desert.

The city of Petra was born 2,500 years ago in a giant bowl hollowed over aeons into the sandstone tableland by wind and water to form a perfect natural stronghold for a desert tribe, the Nabataeans, who lived by raiding caravans on the trade routes that criss-crossed the region. In time they grew wealthy and settled to a more cultured existence, crafting majestic tombs, elegant temples and theatres from the timeless cliffs that enfold the hidden site. They built a sophisticated system of dams, cisterns, pipes and channels to guard it from sudden floods. Later, the Romans added to the city. At its peak Petra may have sheltered 25,000 citizens. Named Rekem in the Dead Sea Scrolls, it had a profound influence on regional culture and politics, but, from the 3rd century on, natural disasters and political tides gradually eclipsed it until it was eventually abandoned and erased from the memory of all but local Bedouin.

In 1812, gleaning rumours of a 'lost city' in the desert, Swiss explorer Johann Ludwig Burckhardt rediscovered Petra and proclaimed its marvels to the European public. English poet John Burgon famously celebrated it as 'a rose-red city, half as old as time'. Approached through a dark and sinuous gorge called the Siq, the site opens out into a breathtaking vista of more than 500 façades of tombs and as many as 3,000 features, over which the Treasury – a royal mausoleum – towers with grandeur. Visitors, particularly in recent decades, have flocked to view these wonders in tens of thousands, providing vital income to Jordan – but also posing a new threat to the ancient city. Added to UNESCO's World Heritage List in 1985, Petra has been listed, by the World Monuments Fund, among the annual 100 most endangered sites four times in the past 12 years.

To gather data for his geo-archaeological information system (GIS), Akasheh and a Bedouin surveyor use a theodolite and measuring rod at the Temple of Qasr Al Bint, dedicated to the main god of the Nabataeans who built Petra more than 2,000 years ago. Wind and water have further sculpted the soft sandstone. Architectural details, state of preservation and degree of erosion of the monuments are included in the GIS being used by Tahani al-Salhi, above left, of the Petra Regional Antiquity Authority.

'Talal Akasheh brings a level of enthusiasm and determination that is rare. He sees that all the various components are implemented.'

DR PATRICIA BIKAI
FORMER ASSOCIATE DIRECTOR, AMERICAN CENTER OF ORIENTAL RESEARCH

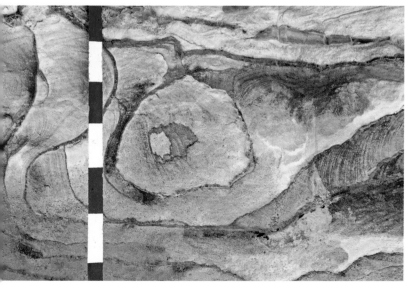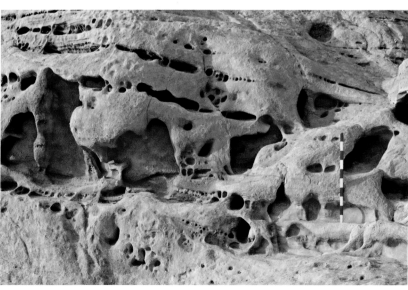

The magnificence of this long-hidden city draws increasing numbers of tourists to the region, both an economic boon and an environmental threat to Petra's survival. They enter through the Siq, below, a narrow defile that leads to the Treasury, left, one of the most elegant of Petra's buildings. Those who choose a camel ride are led along designated routes. Akasheh plans to use the GIS to establish tourist trails for preservation and safety.

From his first visit to the site, Talal Akasheh was not only moved by its beauty – as a scientist, he was eager to understand the chemistry of the weathering that is destroying Petra and to use his training to help arrest it. The vivid colours that lend the city its mystique in fact reflect the chemical processes caused by water within the rock itself: 'Water is the most important element. It gets into the pores of the rock, evaporates and condenses, dissolving the minerals, depositing their crystals which then grow and crack the rock into finer particles. It is a very complex process and the flow of water leaves traces that form these beautiful colours on the surface.'

In 1984, he obtained a UNESCO grant to work with German experts to decipher and remedy these destructive processes. After working for years at the site, he began to absorb the enormity of the challenge – and the many skills it would demand. Early attempts to patch up monuments had failed. Akasheh decided it was vital to document everything known about Petra before proceeding to protect or restore it. Gradually he began to combine the work of archaeologists and geologists, hydrologists, chemists, engineers, architects and planners into a geo-archaeological information system (GIS) and database, with maps and minute details of the site, its physical characteristics, the monuments, their condition and the surrounding modern development.

Today the GIS provides managers of the archaeological park with essential knowledge they need to plan, care for and restore the site and its surrounding region. It offers archaeologists and architects a new way to analyse the monuments and their architecture – and visitors with a safer, more informative experience.

However, the threats facing Petra are pervasive. Akasheh has examined them all: there is the long, slow, chemical disintegration driven by moisture and the airborne salts from the Dead Sea, the rare, fierce downpours, changes to the water table and the sand-blast of desert winds. There is constant abrasion from the hands and feet of tourists and guides, the vandalism of tomb robbers, urban encroachment, the visual canker of modern development and loss of vegetation from the landscape. There is

'No single person or institution can claim to be able to solve the complex problems of Petra. The mere size of the site and the large number of monuments make it a very difficult task.'

TALAL AKASHEH

The sands of time speak volumes in this arid region. A staircase is worn by the tread of man and beast, and a Crusader castle from the 12th century crumbles into ruin. Sandstones in varying colours tell a tale to geologists of desert winds and sudden floods. History, geology and man's presence all have a place in Akasheh's encyclopaedic resource that scientists and experts will mine in the interest of conservation.

the snake-like spread of tarmac roads and the acidic fumes of traffic. 'I had an ulterior motive for wanting to conserve Petra,' Akasheh admits, referring to his fascination with the site. 'Every time I go back there I find new things to marvel at, features to admire. I am myself a tourist. And tourism is the bread and butter of this part of Jordan. But it must be carefully planned. The GIS is the foundation, a first step, in this process.'

Learning the specialist skills to create the database was no simple matter. Meticulously, doggedly, Dr Akasheh applied himself to mastering them. By 2002, the Petra GIS, the first of its kind in Jordan, was in action, acclaimed by colleagues and put to practical use by Jordan's Ministry of Tourism and Antiquities to plan and manage the site. Like the desert itself, funding was erratic, downpours were followed by long periods of scarcity through which the work progressed hand to mouth, often consuming Akasheh's own resources. Yet he persisted. By 2008, the GIS's 10-gigabyte memory collated 2,000 monuments and features, mapped Petra itself, the nearby tourist town of Wadi Musa and the Bedouin settlement of Um Sayhoun. At the same time, he also sought new ways to conserve the monuments and explored the exquisite Nabataean pottery.

Aqel Biltaji, former Jordanian Minister for Tourism and Antiquities, says: 'Dr Akasheh's work is a perfect example of the use of science and technology in the service of sustainable tourism…Jordan lays great importance on tourism development, but is very keen to ensure the protection of the site. Dr Akasheh's work…could be among the first and most effective efforts towards achieving this goal.'

Still, the GIS covers only part of the site – and it is for this vital completion phase that Talal Akasheh has been given a Rolex Award. Over three years, he plans to devote the prize funds to the inclusion of up to 1,000 further possible archaeological features in the database, half within Petra and the rest – possibly including a collection of watchtowers in a rural-suburban agricultural area filled with dams and water-management systems – outside. Study will be made of the flash floods that ravage the site, with a view to possibly restoring the 2,000-year-old Nabataean drainage system. Ground-penetrating radar will be used to prospect the surrounding terrain for hidden tombs and other archaeological mysteries. X-ray fluoroscopy and other advanced techniques will be deployed to study the weathering chemistry of the monuments, to identify those in most urgent need of conservation. The GIS has already yielded a wonderfully detailed tourist map, and may make possible virtual visits to the city from anywhere on earth, via the Internet.

'The area around the city is covered with ancient farm terraces and dams, interesting graves, stone rubble pointing to possible defensive structures on the high ground, and other features,' Akasheh explains. 'If we leave it to the developers, we will never know what was there.'

Among the world's fabled lost cities, Petra's mysterious glory stands as a beacon of human achievement. It has withstood the abrasion of time for over two millennia, but how long it endures depends on how it is cared for today. Eventually, Akasheh acknowledges, it must return to the desert sands: 'Already monuments have disappeared, and some are more affected than others. But it is still worthwhile. It is natural for man to look at his past and to respect it. To want it to last as long as possible. And good documentation of the site keeps its memory safe, even after it is long gone.' |

Gazing into the valley of Petra, Akasheh and a Bedouin companion see the immensity of the challenge – preserving this UNESCO World Heritage Site while accommodating sustainable tourism. So far, he and his associates have documented 2,000 monuments and features. He would like to add 1,000 more. Petra has the largest number of rock-carved monuments in the world. Here man and nature have collaborated in creating this fantastic site. Akasheh will do all he can to prevent its destruction.

Julian Cribb WRITER   Marc Latzel PHOTOGRAPHER

'But from the rock as by magic grown, eternal, silent, beautiful, alone! … match me such marvel save in Eastern clime, a rose-red city half as old as time.'

PETRA, SONNET BY JOHN BURGON, 1845

A Kit to Cut the Smog of Asia's Cities

Reduce pollution from motorized tricycles in Asian cities

Tim Bauer UNITED STATES

In Asia, the ubiquitous motorized tricycle with its two-stroke engine is a major cause of air pollution. Working in the Philippines, American mechanical engineer Tim Bauer and his team have developed a kit to reconfigure these machines, drastically reducing their noxious emissions. He now plans to install thousands of kits, greatly improving air quality and providing tricycle drivers with a way to boost their income.

The best solution is often the simplest solution. Thus Tim Bauer, below right, and his colleagues took off-the-shelf parts, manufactured others and devised an economical retrofit for the millions of two-stroke engine tricycles that crowd the streets of South-East Asia. The outcome: less pollution, increased fuel efficiency and more income for taxi drivers such as Angel Raqueno, below, of Vigan, a UNESCO World Heritage Site in the Philippines threatened by toxic emissions from these vehicles.

Every day at dawn, Angel Raqueno gets on his motorized tricycle with its shaky side-car and criss-crosses the narrow paved streets of Vigan, a small, picturesque tourist city 400 kilometres north of Manila, capital of the Philippines. Long ago, he had begun studying electronics, but he gave it up to take up taxi driving to support his family. For the past 18 years, he has driven passengers through the city's 39 *barangays* (districts) ten hours a day, six days a week. But, for Raqueno, worse than the long hours on the road is the blueish smoke emitted by the 3,000 other tricycles providing transport for tourists and locals around this 16th-century architectural gem: if you are stuck in traffic behind one of these vehicles, the air is almost unbreathable.

According to the *2006 Philippine Environment Monitor* published by the World Bank, atmospheric pollution causes 15,000 deaths in the country every year. Related health costs represent US$19 million a year, and loss of earnings amounts to $134 million. The World Health Organization reports that atmospheric pollution across Asia is responsible for 537,000 deaths a year. The transport sector contributes significantly to this: most of the 100 million tricycles, tuk-tuks, auto-rickshaws and trishaws – symbols of tourism and urban mobility – that clog up Asian cities from New Delhi to Manila are equipped with two-stroke engines – each of them causing as much pollution as 50 cars.

'The retrofit kit introduced and invented by Envirofit and Tim Bauer is a blessing for the city … he is an innovator who has the poor in his heart.'

GLENN CONCEPCION
CITY ENVIRONMENT AND NATURAL RESOURCES OFFICER, CITY OF VIGAN

Easy installation was one of Bauer's goals in developing a direct fuel-injection mechanism for polluting vehicles in Asian cities. His non-profit company, Envirofit, operates in a garage in Vigan, where Bauer and two of his locally trained mechanics, Ruel Regua and Joeman Quiballo, quickly retrofit a motorcycle, a two- to four-hour job. The current kit costs $350, which may be financed with a microcredit scheme, but can be paid for within a year by fuel savings.

Tim Bauer decided that the solution could be found at the heart of the problem. Since 2006, this 31-year-old American mechanical engineer has been distributing a kit that makes it easy to transform the engines on these vehicles into direct fuel-injection mechanisms, thereby reducing the pollution they produce. Carrying out tests in a laboratory and in Filipino garages for many months, it took him and his team every ounce of ingenuity they could muster to disentangle all the technical, economic and socio-cultural intricacies. The result has earned Tim Bauer a Rolex Award.

In the Philippines, about 1.8 million tricycle drivers have to face appalling traffic conditions for long hours every day in order to transport their passengers on congested thoroughfares, roads flooded by torrential rain or riddled with deep potholes. When neither cars nor buses can get through, a tricycle will always find a way. These all-purpose vehicles provide cheap transport for tourists, but also, on a far larger scale, to thousands of people who use them to earn a living, get to work, school, the market or church. 'They play an essential role in the social and economic fabric,' Bauer says. 'But their impact on public health is disastrous.'

In Europe and the United States, two-stroke engines are relegated to powered grass trimmers and chainsaws. But in the Philippines they are used on 94 per cent of motorcycles; in India, Pakistan and Thailand, the figure is between 50 and 90 per cent. For Tim Bauer, it is easy to see why: 'A two-stroke engine is a beautiful thing. It's reliable, robust, powerful and so simple that drivers can repair it themselves, which is very important for people who earn only about $5 daily. But there's a problem: up to 40 per cent of the fuel and oil exit the engine unburned.' This leads to substantial emissions of oxides of carbon, nitrogen and sulphur, hydrocarbons and fine dust, making them one of the main sources of air pollution in the Philippine archipelago.

In 2003, the Philippine government tried to phase out these vehicles and replace them with motorcycles with four-stroke engines, which are

'In recent years air pollution in the cities [in Asia] has been blamed for skyrocketing cases of lung ailments, eye irritations and skin rashes at hospitals and clinics in the region.'

TIM BAUER

less polluting, but cost about $1,500, the equivalent of a tricycle driver's annual income. The authorities were forced to back down when faced with a general outcry from drivers and the vast network of mechanics and sellers of spare parts depending on them. 'The challenge was to find a solution that would allow the drivers to keep their means of subsistence,' says Bauer. 'The constraint was thus to keep the two-stroke and start from it.'

The direct injection kit began to take shape in 2000, in the Engines and Energy Conservation Lab – a spin-off of Colorado State University (USA) directed by Professor Bryan Willson – which Tim Bauer joined in 1997 during his mechanical engineering studies. Bauer, then aged 24, and his colleague Nathan Lorenz were leading a team of students in a research project on the application of direct injection to the snowmobiles of Yellowstone National Park. Bauer immediately saw the potential of this technology for reducing polluting emissions and, at the end of his studies in 2004, instead of applying for a more lucrative job in the aerospace industry, he and Lorenz decided to do their utmost to develop a commercially viable product and make it widely available in Asia.

But getting from North American snowmobiles to tricycles in the Philippines required inventiveness and global awareness. 'I became aware of air pollution at an early age,' Bauer remembers. 'I lived for some time in Saudi Arabia as a kid, and from there I visited Bangkok and Hong Kong with my parents. This is where I saw and felt two-stroke pollution for the first time. It made a lasting impression on me.'

The 'retrofit' consists of a simple but effective mechanical change. 'In a two-stroke, when the piston goes down it uncovers both the exhaust port, where the combustion products are forced, and the fuel and oil intake port. This means that a lot of the oil and fuel mixture is directly washed out in the exhaust stream,' explains Tim Bauer. 'In a direct injection system, fuel is injected into the cylinder later in the cycle, when the exhaust port is closed by the piston, thus greatly reducing the amount of unburned fuel that is allowed to escape.'

The kit can be installed in two to four hours and reduces particulate emissions by roughly 70 per cent and emissions of carbon monoxide

Telltale blue smoke puffs from the exhaust pipe of a taxi in Vigan. Diagrams at right show the differences between the conventional two-stroke engine and a retrofitted one that reduces such pollution. Bauer reports that each tricycle, without the retrofit, produces pollution equivalent to 50 automobiles. Breathing this noxious air leads to 15,000 deaths a year in the Philippines alone.

▌ Conventional two-stroke engine

1. The sparkplug at the top ignites the compressed fuel and air in the upper combustion chamber. The resulting explosion forces the piston downwards. A new charge of fuel and air enter the round crankcase at lower left.

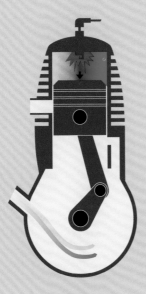

2. When the piston is in the lower position, the new charge of fuel and air enter the combustion chamber through an intake port on the right side. The exhaust port at top left is open to let out gases created by the initial explosion. Part of the new fuel also leaves the combustion chamber unburned, causing pollution in the outside air.

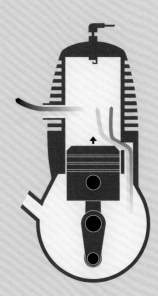

▌ Direct injection engine

1. The critical difference in the direct injection engine is that all fuel is pumped directly into the combustion chamber, top, where both fuel and air are ignited by the sparkplug. The explosion drives the piston downwards.

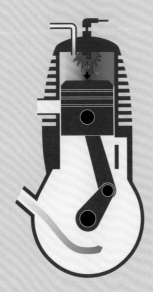

2. When the piston is forced downwards, only fresh air enters through the intake port at right, driving out the combustion gases through the open exhaust port. The piston rises and covers the exhaust port before new fuel is injected into the engine.

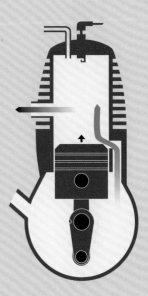

Key

▢ Fuel
▤ Air
▨ Explosion
▉ Combustion gases

Comparison Chart

Red indicates the emissions and fuel consumption from the conventional two-stroke engine. Green shows the greatly reduced emissions after the retrofit kit is installed. The decrease in pollution is shown by the percentages.

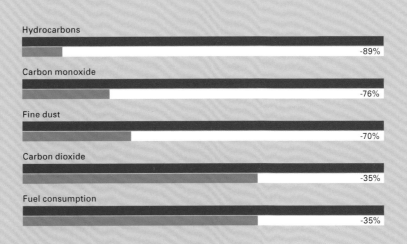

Hydrocarbons -89%

Carbon monoxide -76%

Fine dust -70%

Carbon dioxide -35%

Fuel consumption -35%

'Envirofit has devised the only practical and affordable way…
[to fix] two-stroke vehicles in Asia. This is the here-and-now
solution to go with.'

AMORY LOVINS
FOUNDER, ROCKY MOUNTAIN INSTITUTE

Lined up and waiting for customers, tricycle taxis crowd a street in Manila, capital of the Philippines. Drivers of these vehicles, which provide low-cost transport for locals and tourists, work long hours to earn a living, catching a nap when they can. In Vigan, a popular tourist destination, some taxi owners adorn their vehicles with fancy tin work, hoping to attract riders. One ingenious tricycle owner balanced a towering load of straw goods on his vehicle.

(CO) by 76 per cent, hydrocarbons by 89 per cent and carbon dioxide (CO_2) by 35 per cent. The kit also eliminates the blueish smoke in the exhaust, and oil consumption is reduced by 50 per cent and petrol consumption by 35 per cent – the equivalent of around 450 litres of petrol a year per kit. This makes the engine cleaner than a simple carburetted four-stroke, and for the driver it means a saving of around $3 a day or over $1,000 a year, almost doubling his salary. This extra income is put to use straight away. 'Drivers often give the money to their wife for her to invest – many families have a small convenience store. Or they use it to pay for their children's schooling or studies,' says Tim Bauer.

To keep down the cost of manufacturing the kit – currently $350 – Tim Bauer and Nathan Lorenz used off-the-shelf components: 'We have simply adapted as many components as possible from an existing direct injection system and developed other components (i.e. custom cylinder head, wiring harness, bracketry, etc.) that could be used on the most popular motorcycle models in Asia. One-third of the 30 parts of the kit are produced in the Philippines.'

In October 2003, in order to further develop, commercialize and distribute the kit, Bauer and three of his colleagues founded a non-profit organization, Envirofit, which now has over 20 employees, half of them based in the Philippines. In December 2005, Envirofit signed a Memorandum of Understanding with Vigan City Council, thereby gaining its official support. The following year, it published a troubleshooting manual, translated into Tagalog, one of the main languages of the Philippines, and Ilocano, the language spoken in Vigan. Bauer, who travels to the Philippines five times a year, has also organized about 15 training workshops and seminars in Vigan and Puerto Princesa, two seaside tourist cities with no major industry, where tricycles contribute significantly to atmospheric pollution. Twenty or more drivers and mechanics have attended each workshop so far. 'We have developed the kit so it is easy to install, even by non-certified mechanics,' Bauer explains. 'But we had to convince them that the common idea, according to which the more visible smoke you have, the more powerful your engine is, is wrong. As there is no smoke with the kit, they thought that we were hiding it with some kind of chemicals!'

However, purchasing the kit represents a major investment for a tricycle driver. So Bauer and his team launched a microcredit programme, in collaboration with the Nueva Segovia cooperative bank, which collects repayments on the loans. 'Microcredit is essential to ensure a sustainable impact to our action. Drivers earn money daily, so it's easy for them to pay back their loan and 90 per cent of them do it in less than a year.'

By August 2008, more than 260 drivers in Vigan and Puerto Princesa had fitted their taxis with a kit and had driven a total of over 5.2 million kilometres. With the funds from his Rolex Award, Bauer now wants to further develop the market in these cities and surrounding regions as a stepping stone to distributing the kit more widely in the Philippines and beyond, particularly Pakistan, India, Indonesia and Sri Lanka where millions of auto-rickshaws could easily be retrofitted.

Besides the kit, Vigan City Council is exploring other forms of technology to solve its pollution problem, such as tricycles powered by electricity or natural gas. For the moment, however, their price is prohibitive and, according to Bauer, if implemented incorrectly, can potentially shift the problem elsewhere: 'Two-strokes can have a lifetime of up to 20 or 30 years. If they're banished from the cities, they'll continue to be driven in more disadvantaged, outlying areas. Our retrofit kit makes it possible to reduce the environmental impact of the millions of two-strokes currently in use, and that will still be used for many years.'

Amory Lovins, a world expert on energy resources, agrees: 'Envirofit has devised a practical and affordable way… to fix two-stroke vehicles in Asia. This is the here-and-now solution to go with.'

'These drivers are at the base of the economic pyramid, and these tricycles are a testament to their ingenuity and work ethic,' says Tim Bauer. 'At the end of the day, we can improve their lives with a cylinder head, a few brackets, and, of course, hard work. This is our best reward.' |

Looking forward to a better life, Wilfredo Rabuya rewards Bauer with a smile after the retrofitting of his taxi by Envirofit. The conversion of his and other vehicles will lessen the pollution that attacks Vigan's celebrated Spanish colonial architecture, including St Paul's Metropolitan Cathedral, in the background. Bauer's plans for the future include expanding his programme across Asia.

Francesco Raeli WRITER   Stefan Walter PHOTOGRAPHER

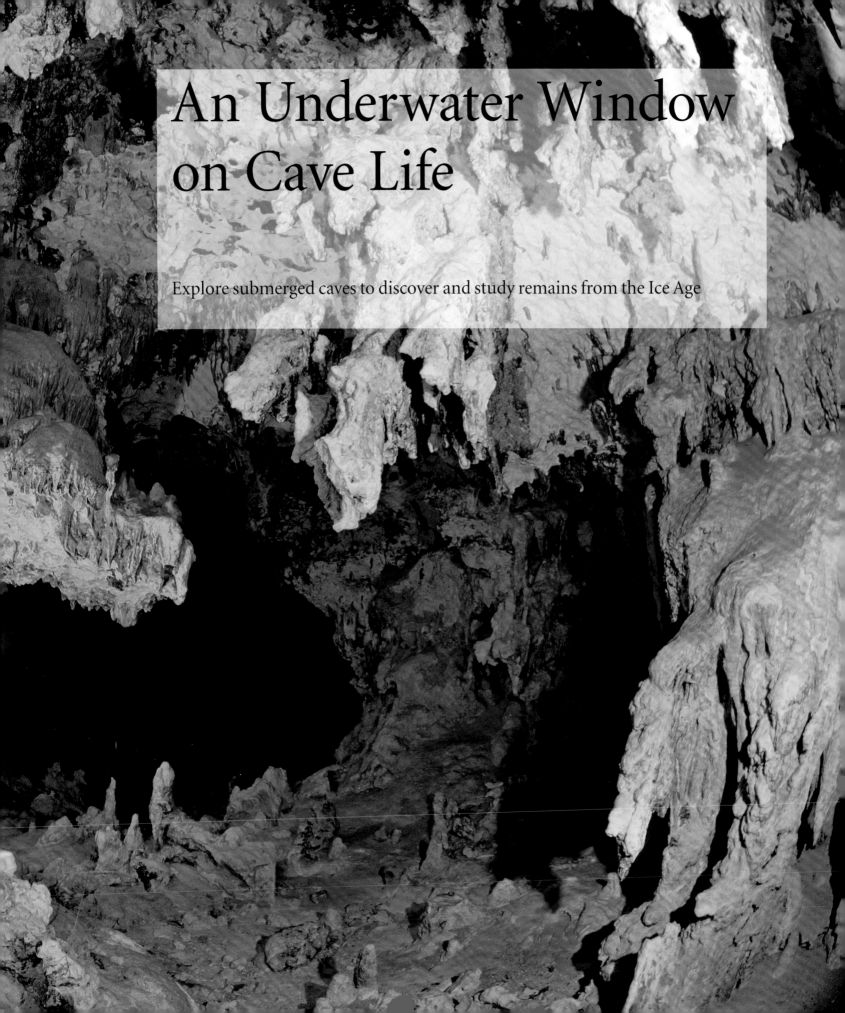

An Underwater Window on Cave Life

Explore submerged caves to discover and study remains from the Ice Age

Arturo González MEXICO

Far beneath the Maya ruins and jungle covering Mexico's Yucatán Peninsula, archaeologist Arturo González has struggled through flooded labyrinths to find proof that humans lived in the region before the end of the last Ice Age sent sea levels rising, inundating the caves. His commitment to bringing the truth to the surface is yielding new insights into prehistoric life.

The serene surface of a *cenote*, or sinkhole, belies the deep and labyrinthine caves that lure Mexican underwater archaeologist Arturo González. The rise in sea level following the Pleistocene epoch flooded ceremonial and burial sites of Ice Age humans that González and his associates have found by arduous diving expeditions into these limestone reservoirs that pock the Yucatán Peninsula.

Mexico's Yucatán Peninsula is a relatively flat landscape where no rivers flow for the rain sinks quickly into the limestone and runs unseen to the sea. The ground is pocked by vine-draped sinkholes – *cenotes*, as they are called locally – where the roofs of underground caverns have collapsed. For centuries these openings have provided inhabitants with access to fresh water, and the inaccessibility of the deep caves beneath the openings has long beckoned the adventurous, though physical challenges limited how far they could go. In recent years, however, technological developments in underwater equipment have made it easier for divers to go farther into the networks of dark tunnels branching out from the submerged caves, and reports began to emerge about this dark underworld and its store of human and animal remains.

Arturo González, a Mexican archaeologist working with the Instituto Nacional de Antropología e Historia, decided to launch a systematic examination of the flooded caverns in 1999. He put together a team of 20 researchers including archaeologists, palaeontologists, divers and photographers who would face technically difficult and physically challenging dives of up to six hours. In the depths, the multidisciplinary team found three human skeletons, and carefully brought them to the surface where they could be preserved and examined. What he found startled the scientific community.

The skeletons are possibly older than any other human remains in the Americas. One in particular has been estimated by three foreign laboratories to be more than 11,600 years old. Furthermore, the skeletons bear no resemblance to the Maya who came to dominate the region thousands of years later, and whose remains and artefacts are found near the openings of the *cenotes*. If anything, according to González, the newly discovered skeletons have a cranial morphology resembling that of people in eastern Asia. The findings are forcing the scientific community to reassess its theories about when and how early humans travelled to the Americas.

'What we've discovered is a piece in the puzzle of human evolution,' says 44-year-old González, who is now director of the Museum of the Desert in the northern Mexican city of Saltillo. 'But there are a lot of other pieces missing from the puzzle. We have one important piece, but it doesn't match any other existing part in a way that would help us understand how early humans colonized the Americas.'

González first learned scuba diving as part of his university studies on biology, but it was a National Geographic documentary about the discovery, by underwater explorer James Coke, of an ancient fireplace 30 metres below the surface that inspired him. 'For me this was unbelievable,' says González. 'Caves have always interested me, this space below the ground that for many indigenous groups signifies the mother's womb. When I saw this documentary about fire pits under the water, I began to travel to these areas to explore them. We got to know James Coke, a pioneer in the exploration of these spaces, and he alerted us to other discoveries he'd made. Thanks to him we began to form a project that since 1999 has been making important discoveries about the ancient history of the Americas.'

Deep in the Cenote Calaveras ('skulls' in Spanish), Arturo González examines two Mayan craniums estimated to be 1,200 years old. He will check for marks that might indicate if they were killed by enemies or priests. The Mayans, who ruled here a thousand years ago, believed the *cenotes* were the home of the rain god, Chac, and in times of drought often offered human sacrifice.

'When you are stuck in a narrow passage, with no visibility, as the minutes go by and you can't move, that's when panic sets in. Then the rule is "stop, think and act".'

ARTURO GONZÁLEZ

'Mr. González' commitment to his work is a quality I have not
seen often. He is a very serious scientist.'

LUIS ALBORES
EDITOR AND LATIN AMERICA DIRECTOR, *NATIONAL GEOGRAPHIC* MAGAZINE

Remains of extinct camelids, giant
armadillos and horses that roamed
10,000 to 60,000 years ago litter
a cave floor, top left. But the most
exciting finds for González and his
team of 20 have been three human
skeletons that are at least 10,500
years old, revising estimates of
human occupation in the Americas.
González demonstrates how to
recover a valuable find, placing
a skull carefully in a plastic box,
above, to carry it to the surface.
He wears gloves and a mask at
the initial examination so as not
to contaminate the prehistoric
evidence.

Deep in the caverns, González and his colleagues have found fossils that are between 10,000 and 60,000 years old, including those of extinct camelids, giant armadillos and horses. All are from the Pleistocene Epoch, when the Yucatán was covered not with low forests but with dry grasslands. In at least one submerged cave north of Tulum, near the Caribbean coast, the explorers found another ancient fireplace, whose carbon traces of partially burned camel bones suggest that the prehistoric humans there survived in part on the meat of an animal that disappeared at the end of the Pleistocene.

When prehistoric people were cooking camel meat, the sea level was more than 100 metres below where it is today. González believes these people may have used the caves not only as rudimentary kitchens, but also as pathways to water sources. There is also strong evidence that dead bodies were placed in special caves far below the ground, perhaps to protect them from natural predators. But then a massive shift in global climate produced rapid rises in the sea level, as well as the intricately linked water table inland, and the water sources, burial sites and kitchens were all flooded – to remain unseen until González's team of divers and underwater anthropologists discovered them.

His findings have greatly increased interest in the *cenotes*, leading González to work with residents of local villages to protect the rare treasures from damage and looting. He has also encouraged them to speak out against the contamination of the underground waters by unrestrained tourist development along the so-called Mayan Riviera.

Funds from his Rolex Award will allow González to field a team for at least another year of research; the group intends to focus on the Chan Hol cave, where a fourth skeleton has been discovered, but not yet removed or analysed. The more skeletons examined, González says, the more comparisons can be made to similar human remains in other parts of the world – perhaps even putting more pieces into the puzzle of human history. Beyond that, González says his team will focus on trying to understand the lives of these ancient people, especially how they used different caves for different purposes – clues that will lead researchers to move beyond the bones and toward a better understanding of prehistoric life.

As knowledge of the past increases, the challenge of getting in and out of the twisting labyrinths remains a highly dangerous pursuit in the name of science and discovery. With complicated logistics and multiple equipment combinations to minimize the risks, the long and disorienting trips underwater remain physically and emotionally gruelling. A typical underwater expedition can take six hours, including the first hour to reach the cave of interest, an hour to carry out research, and then, given the need for decompression stops along the way, a four-hour return trip to the surface. No underwater communication systems have been devised to link cave divers with their colleagues on the surface, so any emergency has to be dealt with by the divers themselves. The wait above can be agonizing. 'I've taken my turn waiting on the surface, worrying all the time, and I prefer to be with the divers below rather than waiting for the team up above,' says González, who will also soon begin doctoral studies at the University of Heidelberg, Germany, focusing on the import of the Yucatán discoveries.

Years of cave diving still lie ahead for González, who is described by Luis Albores, the Latin America director of *National Geographic* magazine, as 'a very serious scientist' whose work is characterized by 'thoroughness and tenacity'. And it's a race against time given the Yucatán's burgeoning tourist development. Yet for González, the risks the divers take as they plunge into the watery windows on the past are worth the challenge.

'As an inhabitant of the Americas, I'm interested in knowing who these people were, where they came from, and when their first steps in the Americas occurred,' he says. 'In these sites, we can find the archaeological contexts just about as they were left by the people of the Ice Age. It's a great opportunity and it is my passion to get there and discover them, and be able to interpret them in order to share a new understanding of the history of humanity.' |

Paul Jeffrey WRITER   Kurt Amsler PHOTOGRAPHER

Changing diving suit for lab coat, González, centre, shows three skulls to Prof. Alejandro Terrazas Mata, left, and Guillermo Acosta, at the National Autonomous University of Mexico. Replicas on a shelf can be examined without damage to the original. González, who studies an animal bone from the Ice Age, will use his Rolex Award to once again explore subterranean caverns in hopes of adding to man's knowledge of his ancestors.

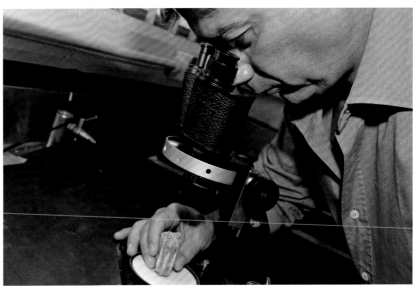

Photographs: Thierry Grobet

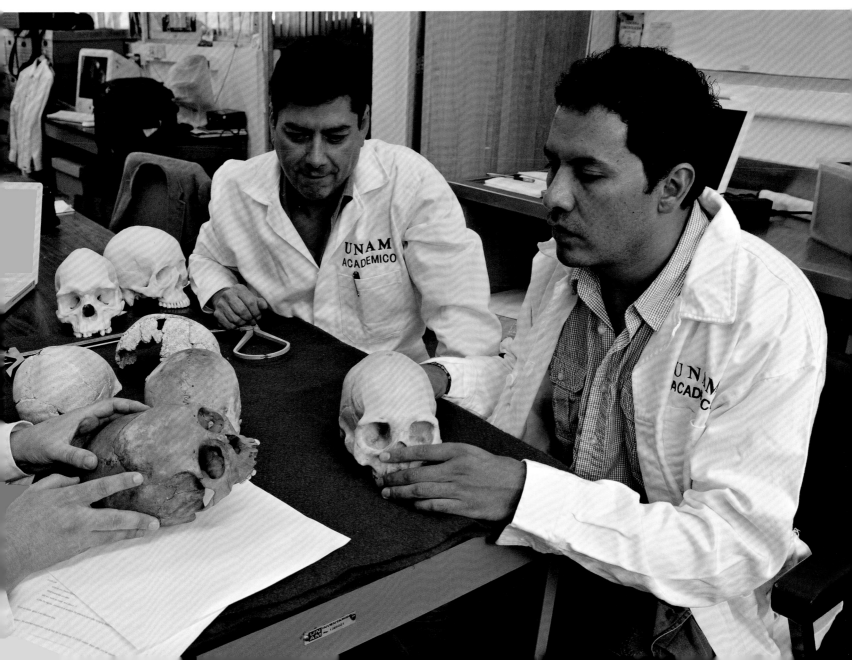

A Kaleidoscope of Life

Safeguard the heritage of the people of Arunachal Pradesh

Moji Riba INDIA

The ancient tribal cultures of India's Arunachal Pradesh state are succumbing to the influences of the modern world. Determined not to let this remote part of the world's heritage slip away, Moji Riba has devised an ambitious plan to involve young people in documenting the state's beliefs and customs and help this rich legacy live on.

Nestled in the Himalayan foothills in the extreme north-east of India is Arunachal Pradesh, an isolated, remote and sparsely populated state that is home to an astonishing diversity of ethnic societies. Few regions of the world can match the wide range of languages and religion, diet and dress enclosed within the state's 83,700 square kilometres. A million inhabitants are divided into 26 major tribal communities, each with its own distinctive dialect, lifestyle, faith, traditional practices and social mores, living side by side with about 30 smaller communities. In the far west live the Sherdukpen and Monpa tribes, practitioners of the ritualistic Tibetan form of Buddhism, adept at mask making and pantomime dances. The Adi and Nyishi people live in the state's heartland and worship their gods at elaborate altars crafted out of bamboo and cane. In the extreme south-east of the state, the Wanchos are known for the quality of wood carvings they create. In the east of the region, the Idus are expert textile weavers and their costumes are a rich tapestry of hues and designs, while the Apatani are famous for their basket weaving, their strong village institutions and social networks. The Simong, who inhabit the forbidding northern uplands, climb up mountain peaks to perform their rituals and collect poisonous aconite plants to use on their arrow tips for hunting.

For 36-year-old film-maker Moji Riba, the cultural richness of Arunachal, his home state, is 'like a wonderful shawl woven in a myriad of colours and patterns'. Situated far east of the bulk of India, the state has borders with Tibet (China), Myanmar and Bhutan, but it is isolated by high mountains and dense forests, and regulated by a strict tribal protection policy that requires even Indian citizens to have a special permit to enter the region. As a result, the ethnic groups of Arunachal were, until recently, shielded from external influence. 'The Arunachali have evolved an enviable understanding of their immediate environment, finding imaginative ways of survival in their rugged homeland,' Riba

Culture is a two-way street in Tawang, in India's far north-east state of Arunachal Pradesh. Young Buddhist monks in their crimson cloaks want a look at the cameras of film-maker Moji Riba, who is documenting the folklore and diversity of the many tribes of the region. At the Tawang school, the correct answer to a question from Riba about tribal ways earns a student a free notebook.

'Moji Riba's knowledge comes both from inside the tribe and from the wider spectrum of his experience of the rest of the world.'

DR SHOBITA PUNJA
INDIAN NATIONAL TRUST FOR ART AND CULTURAL HERITAGE

An avid pupil, Hage Komo gets video instruction from Riba, who is enlisting local young people to capture the oral histories, languages and rituals of their tribes for his Mountain Eye Project. Komo films his father gathering bamboo in a grove outside Hari Village. In the village rice fields, Landi Rinyo of the Apatani tribe plants millet on raised dry mounds in the traditional way.

explains. 'Over time, they have devised a bold celebration of the pageantry and patterns of everyday life. There is much to learn from their contributions to folklore, arts and crafts and philosophy.'

Today, however, economic development, improved means of communication, the exodus of the young and the gradual renunciation of animist beliefs for mainstream religions threaten Arunachal's colourful traditions. 'It is not my place to denounce this change or to counter it,' says Riba. 'But, as the older generation holds the last link to the storehouse of indigenous knowledge systems, we are at risk of losing out on an entire value system, and very soon.' The risk of many of these cultures disappearing in a generation is particularly great as almost the entire body of local wisdom – from religious chants to tribe histories, from love songs to agricultural rituals – exists today only in the oral tradition. The death of every older person in a village means the loss of part of the local heritage.

This realization led Riba, who holds a master's degree in mass communications, to set up the Centre for Cultural Research & Documentation (CCRD) in 1997 in Itanagar, the capital city of Arunachal. Over the past decade, a team working at the centre has made 35 documentaries for national television stations and for government and non-governmental agencies. But the centre is more than just an archive or library, it is also a

platform offering the tribal people an opportunity to voice their concerns and share experiences. In 2004, Riba was instrumental in creating the diploma in mass communications at Itanagar's Rajiv Gandhi University, again to augment understanding of cultural values and local customs. He currently juggles his time as head of the university's communications department with running the centre.

But the unprecedented rate at which cultural change is taking place is simply 'too large, too rapid and too overwhelming' for the team to capture through standard methods. Riba's solution is the Mountain Eye Project, an unconventional and ambitious initiative that aims to create a cinematic time capsule documenting a year in the life of 15 different ethnic groups. Riba will select and train young people from each community to do the filming. This gives him access to enough film-makers – a resource he lacked at the cultural research centre – as well as access to people with an intimate understanding of village life. Beginning in early 2009, these novice film-makers will capture a broad range of the tribes' oral histories, as well as the rituals, ceremonies and festivals that take place over a year in their villages. Riba expects to collect about 300 hours of film per village, all of which will be recorded and archived in the native languages. He believes that the resulting 4,000-plus hours of video will provide an

Map of Arunachal Pradesh

Map not drawn to scale

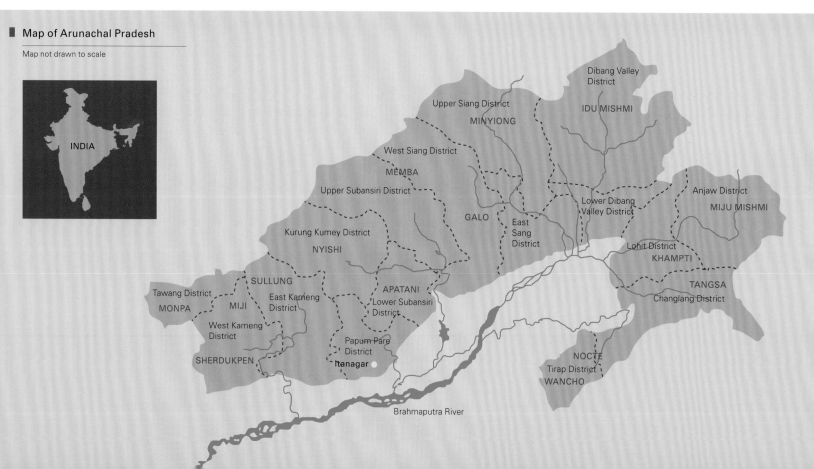

INDIA

Dibang Valley District

Upper Siang District

IDU MISHMI

MINYIONG

West Siang District

MEMBA

Upper Subansiri District

GALO

East Sang District

Lower Dibang Valley District

Anjaw District

MIJU MISHMI

Kurung Kumey District

NYISHI

Lohit District

KHAMPTI

SULLUNG

APATANI

TANGSA

Tawang District

East Kameng District

Lower Subansiri District

Changlang District

MONPA

MIJI

West Kameng District

Papum Pare District

SHERDUKPEN

Itanagar

NOCTE

Tirap District

WANCHO

Brahmaputra River

'What has emerged … is the recognition that the rich cultural
heritage of the proud people of this state … faces the danger
of completely disappearing.'

MOJI RIBA

Dressed in their finery in the village
of Yazali, Nyishi women dance
to honour their animistic religion,
Donyi Polo, which means 'sun-
moon'. Riba's project will focus
on 15 selected tribes, shown in
red, in this remote region. Long
isolated by an Indian tribal
protection policy, the people are
now exposed to myriad outside
forces including development
and globalization.

invaluable record of life as it has been lived in his state for centuries. The
project will also engage scholars belonging to the 15 tribes from the
University at Itanagar to analyse and translate this vast amount of data
and organize it in a publicly accessible database.

Riba is philosophical about the inevitable limits on what can be saved
for the future. 'While there is unquestionably a need for the documentation of customs and beliefs,' he says, 'we also understand that all this
documentation and the outreach activities will not ensure that these
customs continue to be practised in their original forms – it would be
unrealistic to even expect it to do so. The forces of change are larger than
what we can take on. We are instead trying to create a space where they
will continue to live in some form or other: some definitely in their prac-
tice like the singing of songs, the telling of folk tales and the fun of the folk
dances; others, like the Apatani nose piercing or the Wancho tattoos and
the war rituals, to be understood and valued for what they would
have meant to our people in another time and age. I like to think of our
heritage as an elastic band. I want to stretch this as much into the future
generations as we can – till it reaches its edge and snaps. Each day I wake
up and hope that this never happens. But that is sadly a finality we have to
stare at – unless of course, there is a revolution of some kind!'

Each video volunteer will also produce a film based on a particular
aspect of their village that moved them. These short films will be the
cornerstone of an outreach programme, to be held initially at Itanagar's
Jawaharlal Nehru State Museum. A series of interactive workshops will
take place where the films will be shown alongside traditional artistic
activities such as mask making, painting, storytelling and participatory
games using local languages. Cultural heritage activity clubs will be
launched in participating schools and colleges. Students will be encouraged to create information posters about various tribes and undertake
field trips to the State Museum. Schools will hold events where students
from different tribes can meet and exchange cultural experiences.

'It is my hope these outreach activities will inculcate in children and
youth an appreciation of their traditional heritage, and help them make
sense of their ancestry, their identity,' reasons Riba. The films will then be
screened in New Delhi, giving the novice film-makers an opportunity to
present their work to a large audience in a country where cinema is one of
the most popular forms of entertainment.

'This is a path-breaking project, for it views the folklore and cultural heritage of the tribal groups … not as mummified objects to be confined to museums but as a living thing.'

PROFESSOR K. C. BELLIAPPA
RAJIV GANDHI UNIVERSITY, ARUNACHAL PRADESH

Ultimately, Riba hopes to draw attention to this part of the world, enhance the centre's visibility and encourage support from other sources. To get the project off the ground, he needs to purchase video equipment – each video volunteer will be equipped with a complete digital video documentation unit – and fund training. He explains that the funds from his Rolex Award will be used as seed money, launching the project and encouraging other donors to give support. 'The scale of the project is large, it's almost like working against the clock to try and get as much done within a limited time frame in a vast area. Therefore, the resources needed are also relatively large. There is a crying need to fill the vacuum that exists in providing a platform for issues like the promotion of the indigenous languages, the ideal of tribal identities as a common shared heritage and the use of heritage education to enable the future generations to share, realize and respect the diversities in culture,' says Riba.

Professor Kambeyanda Belliappa, of Rajiv Gandhi University, needs no convincing about the far-reaching implications of Mountain Eye: 'This is a path-breaking project, for it views the folklore and cultural heritage of the tribal groups in Arunachal Pradesh not as mummified objects to be confined to museums, but as a living thing that needs not only to be documented but also passed on to the next generation.'

Riba is naturally eager to preserve his own family's heritage (part of the Galo ethnic group), as well as the many other cultures of Arunachal Pradesh. Like many young Arunachali, he speaks – and thinks – in English, deemed the language of opportunity in India, and also the language of instruction at his school and at university. But he vividly recalls his father's funeral in 2000, where he felt more like a bystander than a participant while his relatives embraced his father's body and sang a lament for him, an *ane-naenaam*. It pained him that he could not understand the eulogies they were offering his father or the memories of him that were being shared. 'I could not even thank the people that sang them,' he concedes. 'Indigenous languages have been caught in the crossfire between English and Hindi, the national language of India. Today, I am making a concerted effort to learn my mother tongue, the Galo language, and to encourage my boys – nine-year-old Jiri and Jili, aged three – also to learn it. Language is a significant part of our culture, our heritage, and we cannot afford to let it die. In today's era of globalization, where everybody is encouraged to be the same as everybody else, language is one of the only things we have left to distinguish ourselves. Mountain Eye will help preserve these languages and hopefully encourage the audience to bridge the divide between modern society and their tribal identity, inspiring them to be in touch with their roots.' |

A mosaic of faces, colourful dress and handcrafted adornment reveals the tribal variety of Arunachal Pradesh. Clockwise from top left: Wancho, Galo, Khampti, Apatani, Idu Mishmi, Monpa, Nyishi, Nocte and Idu Mishmi, at centre. The rich traditions of these tribes and more will be preserved in Moji Riba's Centre for Cultural Research and Documentation, a treasure trove to be mined by future generations.

Alexa Schoof Marketos WRITER   Xavier Lecoultre PHOTOGRAPHER

Maligned Heroes of the Mexican Night

Save endemic and endangered bats through habitat protection and education

Rodrigo Medellín MEXICO

Superior pollinators and insect predators, bats are, with a few exceptions, valuable assets to mankind. Yet these flying mammals are reviled and killed. Rodrigo Medellín's passionate advocacy of bats through conservation and education is dispelling harmful myths and bringing about harmony between these animals and their human neighbours.

Once worshipped as deities, bats held a place of honour in the rich cultural landscape of the Maya civilization. But these remarkable animals – in some areas of the world a keystone species critical for healthy ecosystems – have suffered from centuries of misconceptions and folklore that portray them as sinister, disease-carrying, blood-sucking demons. Out of ignorance and fear, humans wipe out entire colonies of bats through pesticides, encroachment or by mining, burning or dynamiting the caves where they roost.

Severely under-studied, omitted from conservation plans, bats are among the world's most rapidly declining mammal species. A total of 1,116 species of bats exist worldwide, and are found everywhere except polar regions and desert extremes. Eighty-five species are endangered, and in many cases, the main threat to them stems from mankind's fear and hatred.

For Rodrigo Medellín, however, bats are nothing less than astonishing. Mexico's foremost authority on bats and an ardent conservationist determined to change their image, he is Professor of Ecology at the National Autonomous University of Mexico (UNAM), where he has devoted over 30 years to creating awareness of the invaluable role they play in keeping ecosystems and lucrative agricultural crops healthy.

Medellín was 12 years old when he first encountered bats in a hot, damp cavern teeming with life. Vampire bats hung in one corner and nectar-feeding bats mated in another. Insects burrowed into mounds of bat guano, while a snake hunted sleeping bats. 'It was incredible. Surrounded by life, I couldn't find a single spot on which to focus,' he says. Discovering this wealth of biodiversity in that single location was a pivotal moment in his decision to study one of the most ecologically diverse mammals in the world and to correct the many misconceptions about them.

The scientist waits for dusk in the Sonoran Desert when adult lesser long-nosed bats will fly from Pinacate Cave to feed on and pollinate agave plants and saguaro cactus flowers. Then Rodrigo Medellín, Mexican biologist and conservationist, will enter the cave to study their two-week-old babies. In an infrared photograph, previous pages, they cluster on the ceiling.

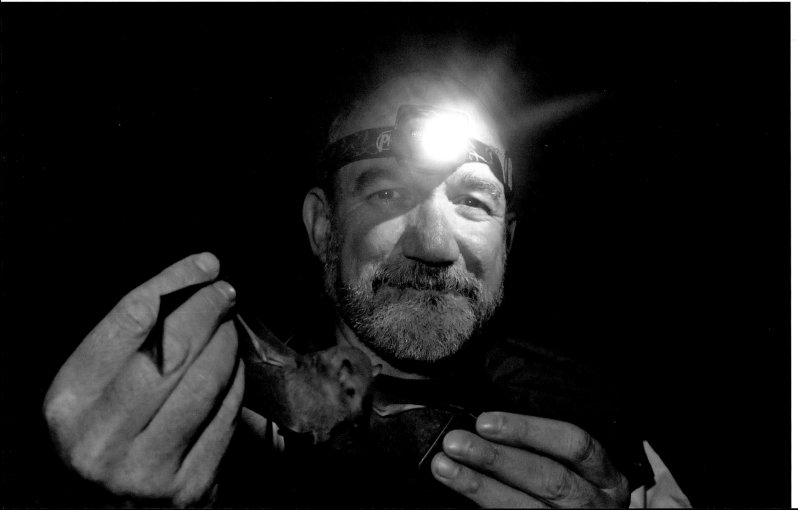

Bats come in incredible variety. Mexico alone has 138 species and 19 of them are threatened or endangered. The latter include the little-known banana bat, opposite top left, the flat-headed bat, above, believed extinct until Medellín rediscovered it in 2004, and the Baja California fishing bat. He spreads the wings of a Mexican long-tongued bat, so named because its tongue can extend one-third of its body length.

Bats are natural controllers of night-flying insect pests and consume almost the equivalent of their weight in mosquitoes and crop pests each night. Corn earworm moths cost farmers billions of dollars annually, yet in one night, a million Mexican free-tailed bats (*Tadarida brasiliensis*) can destroy 10 tonnes of moths.

Across Mexico's lush rainforests, sprawling savannahs and vast deserts, bats pollinate flowers of many hundreds of species, such as columnar cacti and agaves (a vital ingredient in the production of tequila), and disperse seeds of many species that promote forest restoration. In fact, bats distribute up to five times more seeds per square metre than birds, and can account for up to 95 per cent of forest regrowth. Mexico, renowned for its extraordinary biological diversity, boasts an astonishing variety of bats – Medellín estimates that his country has 138 species, of which 19 are officially threatened or endangered.

Three of Mexico's bat species feed on the blood of higher vertebrates, and there have been reports of the common vampire bat, *Desmodus rotundus*, attacking humans. But, Medellín says, 'in Mexico, this is a very rare occurrence. The risks to humans are extremely low. In fact, virtually all bats are completely harmless, and 100 per cent beneficial, even crucial to ecosystems and humans.'

With the benefits brought by so many bat species so high, and the risks to humans from a handful of species so low, Medellín saw the significant decline in the population of his country's 10 major bat colonies as a call to action. In 1994, he founded the Program for the Conservation of Bats of Mexico, in partnership with his university and Bat Conservation International. Under his direction, a comprehensive strategy was established, based on research, education and conservation.

'This project really makes a difference. The education component is wonderful. If he can convince me that bats are good, he can convince anybody!'

DR RODRIGO JORDÁN
EDUCATOR AND CHAIRMAN OF CHILEAN NATIONAL FOUNDATION FOR THE ALLEVIATION OF POVERTY

'Rodrigo Medellín's distinguished career in research has been complemented by his outstanding, longstanding dedication to outreach and education.'

DR ELEANOR STERLING
DIRECTOR, CENTER FOR BIODIVERSITY AND CONSERVATION, NEW YORK

Today, Medellín and his 30-member team, drawn from master's and Ph.D. students, identify priority sites among Mexico's estimated 30,000 caves, and then develop management and recovery programmes for threatened species. Ranchers, for instance, believing them to be vampire bats that prey on their cattle mistakenly destroy thousands of beneficial bats; Medellín and his team defuse the problem by teaching them vampire-bat control strategies. As part of the programme, educational materials are made available, community workshops held regularly and an accurate picture of bats and their usefulness is presented via nationwide media exposure, including an award-winning radio show that reaches millions of listeners. 'Adventures in Flight' is a series of 15-minute broadcasts, aimed mainly at children, with each programme covering an aspect of bat biology or conservation.

Medellín's overall strategy has proved highly effective, becoming a model for similar initiatives in Bolivia, Costa Rica, Guatemala and elsewhere. He believes that if young people do not change their attitudes, bats are doomed. His teams work with schools and communities located near habitats for threatened bat species. Games, toys and storybooks are used to enlighten children. 'We've reached well over 200,000 people, at least half of them children,' Medellín says, adding that thanks to radio programmes, coverage on television and articles in the press, millions of people now have access to accurate information about bats.

One striking example of the strategy's success occurred in 1996, soon after Medellín and his team worked with a school near Monterrey in northern Mexico. Rumours began to circulate that a livestock-killing creature, the Chupacabras, lurked in the famous Cueva de la Boca caves, home to the world's largest Mexican free-tailed bat population. Locals threatened to burn the cave until schoolchildren – newly informed about bats by Medellín's team – intervened and explained their benefits. Local people grew to appreciate the Mexican free-tailed bat, and its population increased from a low of 100,000 in 1991 to 2.5 million by 2001. 'To this day, the cave remains protected and cherished,' says Medellín.

The answer to fear is knowledge. Medellín provided children at an elementary school with stuffed specimens of the world's only flying mammal, hoping familiarity breeds empathy. A schoolyard game helps explain how bats pollinate. Students with spoons are bats, others play the flowers. Medellín, who has been called 'the Rock Star of Bat Conservation', performs at a secondary school in Mexico City with a picture of a big-eared bat projected on the wall.

He explains that one major challenge of his work is convincing those funding conservation that bats are worth supporting. 'This is a constant battle because most donors focus on charismatic species such as big carnivores or birds. Patience and education are needed, we have to explain to donors the importance of investing in bat conservation.'

The funds from Medellín's Rolex Award will be a welcome boost, allowing him and his team to work in ten states, selecting ten new priority caves beyond the 25 that had been previously identified as needing conservation. They will also focus on five endangered species, including the flat-headed bat (*Myotis planiceps*). Declared extinct by the World Conservation Union (IUCN) in 1996, this tiny animal – at 3 grams, one of the world's smallest bats – was rediscovered by Medellín and his associates in 2004.

Deeply committed to safeguarding not only bats but all of Mexico's wildlife, Medellín is extending his work to other species, including the pronghorn antelope, bighorn sheep, black bears and the first-ever nationwide population estimates of jaguars in Mexico. In demand at conferences and universities worldwide as a speaker and educator, Medellín has become a potent force in changing negative perceptions and restoring pride in one of Mexico's most unusual animals, earning along the way several major honours, including the Whitley Award.

'Rodrigo is brilliant, and … because of his intellect, passion, commitment and humour, he is able to convince people from all walks of life of the importance of bat populations, and their need to get informed and involved in their conservation,' says Dr Mary C. Pearl, president of the Wildlife Trust, in New York.

For Medellín, the words of a young boy are the best validation of his work. A few years ago, he says, 'after my education team had already worked in a cave in western Mexico, I arrived at the cave incognito with some donors. As we got out of our vehicles, a child no older than nine approached us and offered to tell us about the importance of the bats that lived in that cave if we gave him a peso. I immediately gave him a couple of coins and he proceeded to tell us all about bats and their pest control, pollination, and seed dispersal services. I could not have been happier!' |

A flurry of threatened lesser long-nosed bats swirls about the head of one of Medellín's students in Xoxafi Cave, one of 25 that he has identified as crucial for this species' recovery. The community that owns this cavern educates visitors about the benefits of bats: pollinating plants, dropping seeds that aid reforestation and eating insects harmful to man and agriculture. The spread wings of a captured bat seem almost transparent in the glare of a spotlight.

Lynne Schuyler WRITER   Thierry Grobet PHOTOGRAPHER

'New data will justify the creation of protected areas for bats. Society is catching up rapidly on the importance of bats for humanity's well-being.'

RODRIGO MEDELLÍN

An Inedible Waste to Prepare Billions of Family Dinners

Turn rice husks into cheap, clean energy for cooking

Alexis Belonio PHILIPPINES

His fertile mind teeming with inventions to make life better and save energy, Alexis Belonio has produced a simple, gas-fired stove powered by rice husks, one of Asia's most abundant farm wastes. He is ready to share his low-cost invention, which reduces fuel costs and minimizes greenhouse gas emissions, with millions of families in the Philippines and abroad.

112 | 113 The clear, blue flame emanating from the burner of the humble, metal cooking stove is a luminous sign of hope for hundreds of millions of poor farming families. And yet inventor Alexis Belonio was initially blithely unaware that he had achieved what leading experts had declared could not be done: turn agricultural waste into purified gas for domestic cooking in a top-lit, updraft, biomass gas stove.

In the many countries where rice is the staple food, small changes to the use of this grain can make a profound difference. The world's annual 650-million-tonne rice crop provides sustenance three times a day for 2 billion people, mainly in developing countries in the tropics. For these people, this staff of life is indispensable, but to cook the rice they need gas or kerosene, fossil fuels that come at an increasingly unaffordable cost and with often negative consequences on health and climate.

But there is another side to rice: the huge piles of inedible husks that are often found rotting beside roads or smouldering in fields, millions of tonnes of potential energy, mostly going to waste. For Belonio, a 48-year-old associate professor of agricultural engineering in the Philippines and inventor of over 30 devices to help farmers, many of them poor, finding uses for this neglected abundance became an obsession.

Surprising the experts, Alexis Belonio built the first efficient and economical gas stove using rice husks, which are usually dumped along the roadside, preceding pages. He proved the secret lies in converting the husks to gas which burns with a hot, blue flame, right. In a workshop Belonio carefully measures the diameter of the burner rim.

Cookers fuelled by rice husks have been used before, but they are sooty and unhealthy; nor can they generate enough heat to cook food quickly. Belonio believed that if he could convert the rice husks to gas, it would provide a much hotter, cleaner flame to cook on. Gasification has been regularly reinvented for many purposes over the past 150 years, including for several types of stoves, but few applications have promised to benefit so many people, so simply and so cheaply.

Drawing the concept from a technical workshop on wood gasification at the Asian Institute of Technology, in Thailand, then working alone and with his own resources, he designed a simple, top-lit stove with a small fan at the base supplying an updraft of air. In Belonio's design, a stream of oxygen converts the burning rice husk fuel to a combustible blend of hydrogen, carbon monoxide and methane gases, yielding a hot, blue flame similar to that produced by burning natural gas. At first, says Emeritus Professor Paul S. Anderson, of Illinois State University, Belonio 'was unaware that what he was trying to do had been deemed ... as not being possible. He did not even know he should have been highly surprised that he succeeded!' Belonio says simply: 'It is a God-given technology. I wish to share it with people all around the world.'

114 | 115

An old hand at using the new stove, Emok Galeo has been cooking on one for three years. The stove consumes a kilogram of rice husk per load (see diagram at right), providing 40 to 50 minutes' cooking time. The unit releases neither smoke nor toxic fumes and fries fish in minutes.

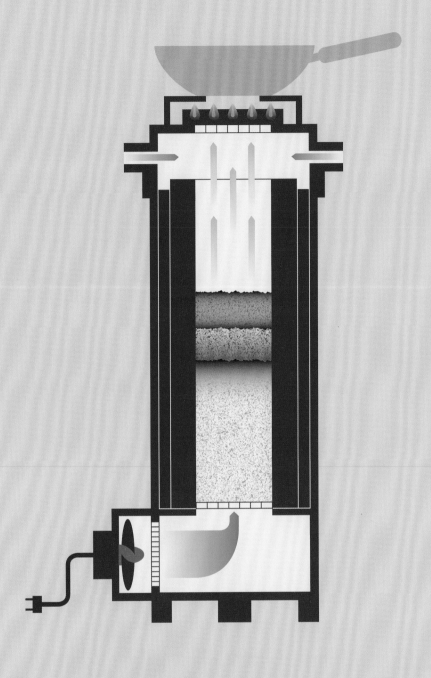

Key

- ■ Rice husk
- ■ Air
- ■ Burning layer
- ■ Char
- ■ Combustible gases

Belonio's Rice Burner

Simple and effective, the process begins by filling the rice burner's main chamber with husks and lighting them. Then a fan introduces air that feeds the burning husks. This partial combustion produces char or carbon that reacts with oxygen in the air to generate combustible gases such as carbon monoxide (CO), hydrogen (H_2) and methane (CH_4). The combustion of these gases, helped by additional air entering through the burner holes at the top, produces a blueish flame.

'The world produces over 115 million metric tonnes of rice husks each year. Using them for cooking and other related applications can save a great deal of energy.'

ALEXIS BELONIO

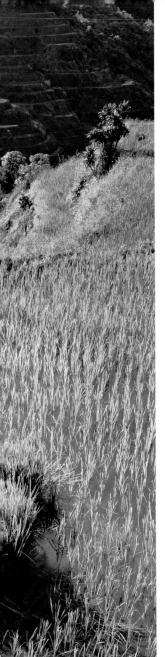

Photograph: Stefan Walter

The Philippines' premier crop, rice is grown throughout the islands – from the steep terraces of Luzon to southern Mindanao. More than 2 million hectares are devoted to rice planting. At the Facoma Rice Mill on Panay Island, the husks removed from the kernel grow into a large pile. A travelling rice mill serves small farmers who take their hulled rice home in metal boxes.

But there was a setback. Belonio's early stoves, made in the Philippines, sold at US$100, too expensive for a poor family to afford. Further research and development conducted in Indonesia significantly reduced the retail price of the stove to only $25. This was achieved by simplifying the design of the stove in terms of operation, materials and fabrication. Thousands of cookers are now being manufactured by companies cooperating with Belonio in the Philippines, Indonesia and Cambodia. By exploiting a freely available waste product at a time of soaring energy prices, the stoves can save a family of rice farmers about $150 a year in fuel bills, a huge benefit for families that live on $2 or $3 a day.

The potential benefits are prodigious. A tonne of rice husks contains the same energy as 415 litres of petrol or 378 litres of kerosene. A few handfuls of rice husks can boil water in six to nine minutes. Best of all, the husks are usually free, either on the farm or from the waste dumps that surround rice mills. Furthermore, by being far more efficient than ordinary cookers, Belonio's stoves reduce greenhouse gas emissions and eliminate toxic fumes inside houses. Even the char left after burning can be recycled to improve farm soils or to form bio-coal briquettes.

Belonio has recently scaled up the principle of his domestic stove to create a whole family of new technologies: dual-reactor and continuous-flow gasifiers for grain dryers, bakery ovens, commercial kitchen stoves, and small power-generating plants. His latest invention, a 'super-gasifier', is a powerful rice-husk stove driven by injecting superheated steam, which, he says, is ideal for cottage industries. 'I was very surprised at how well it worked. That was a great moment.' His technologies are proven, reliable and inexpensive. In addition to rice husks, they can use other biomass such as coconut husk, corn cobs and sugarcane bagasse, instead of fossil fuels or timber from fast-vanishing rainforests.

Bags upon bags of rice fill a room at the Abordo-Portaje rice mill in Janiuay on Panay Island. Just outside the mill, Belonio, bottom, has helped the owner retrofit large paddy furnaces to burn rice husks, realizing a saving of about US$8.00 an hour for fuel. Mr Abordo uses the furnaces as a heat source for a dryer and the rice hull char becomes soil conditioner for his rice farm.

'Many eminent people in the field did not consider Belonio's research in gasifiers to be possible, but he has made it work.'

DAVID HELMER
VOLUNTEER, CANADIAN EXECUTIVE SERVICE ORGANIZATION, PHILIPPINES

Belonio's ambition now is to spread the word about his inventions and to share the know-how, in the Philippines and around the world. He has already published a handbook on building the rice-husk gas stove, which is available for free on the World Wide Web. With funds from his Rolex Award, Belonio plans to set up a demonstration centre in Iloilo, in the Philippines, to disseminate free information and to provide training and technical advice. He will also research new inventions, such as a large-scale, rice-husk-fuelled gasifier and a gas-turbine power-generating plant for supplying electrical energy to rice mills and for lighting remote villages. He even envisions storing gas from rice husks to run farm machinery.

Professor Anderson says of this keen inventor: 'Alexis' accomplishments are founded upon his personal drive and the use of his personal resources. Establishing a centre dealing with rice husks is a worthy goal that will eventually benefit millions of people in many countries.' With tireless dedication, practical focus and technical insight, Alexis Belonio is continuing his quest to convert overlooked energy sources to ensure that families in many developing countries can prepare their meals more cleanly and efficiently, at the least cost possible to the environment and to themselves. |

Julian Cribb WRITER   Kirsten Holst PHOTOGRAPHER

An Unexpected Protector for India's Rainforests

Establish a network of rainforest research stations across India

A first love, snakes still delight Rom Whitaker, noted herpetologist and conservationist. At the age of five he captured a milk snake at his home in New York state. Now a citizen of India, he hopes to help save the threatened rainforests of his adopted country by establishing a network of research stations similar to his headquarters in Agumbe, where he shows volunteers a vine snake.

Romulus Whitaker INDIA

Over the course of his unconventional life – from being an American boy raised in India's wild places to achieving scientific renown – Romulus Whitaker has gone from conserving reptiles to saving rainforests. His planned network of research stations will build on the surprisingly limited knowledge of India's rainforests and demonstrate the importance of their water supplies to hundreds of millions of people.

Romulus Whitaker seems an unlikely name for an Indian citizen who is also a reptile expert and environment film-maker. But the combination of a foreign name, mildly Viking looks inherited from his Swedish mother, an unexpected fluency in local Indian dialects and a thoroughly irreverent attitude has marked this American-born, 65-year-old Indian citizen out as a highly unconventional yet effective conservationist in a country far from his birthplace.

Whatever his ancestry and skills, what drives him is a boundless enthusiasm for the wonders of nature, and a determination to save them. 'It is fascination with the endless natural mysteries, questions on why critters do what they do and empathy and sympathy in the face of the destruction all around,' he explains from his base in the southern Indian state of Tamil Nadu. 'I haven't had to do a nine-to-five job ever in my life, and that is a very envious situation to be in if you like the wild. Life has been much like a river in that it picks you up and carries you along. I have got into things as they come towards me.'

This seemingly relaxed attitude belies the original thinking and careful and considered planning behind his many projects for wildlife, for forests and for the people living in them. His current ambition, for which he has been selected as an Associate Laureate in the 2008 Rolex Awards, is to create a network of rainforest research stations throughout India, part of a vision he has been elaborating in his mind for many years. 'The idea of the rainforest research stations has been with me absolutely forever, but I didn't have the wherewithal to do anything about it. Then all these things started falling into place over the last few years. My mother died and she left some money, enough to buy this block of land [at Agumbe, in southern India] we had talked about before her death. Then the Whitley Award for Nature came

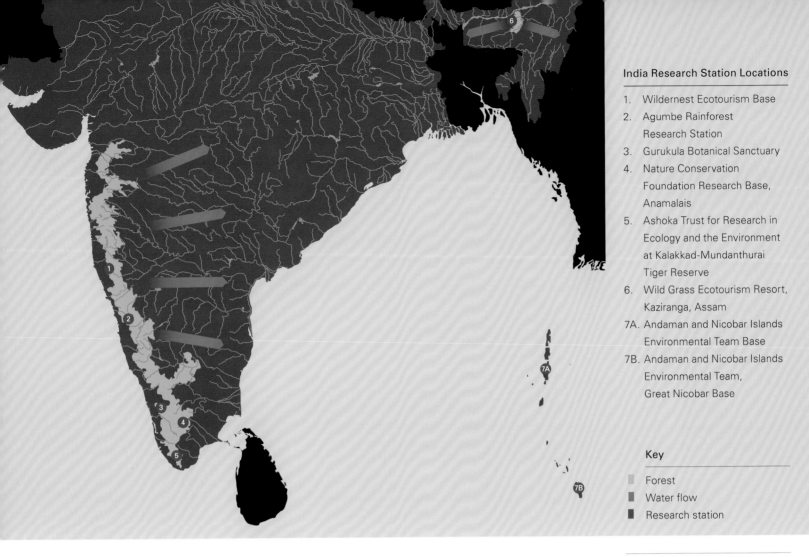

Key

■ Forest
■ Water flow
■ Research station

along and helped set up the Agumbe Rainforest Research Station and get it working really well.'

A mother's tolerance for a small boy's fascination with snakes – first in upstate New York and then among India's much more venomous varieties – became the basis of a notable career in herpetology for Whitaker. Author of eight books and over 150 articles, he served in key reptile posts and has inspired many with 23 acclaimed environmental documentaries, such as the National Geographic film *King Cobra*. In 1984, for his project to help the indigenous Irula people of Tamil Nadu make the transition from their old trade (catching snakes for the now-banned skin trade) to collecting snake venom to produce life-saving anti-venom serum, he received an Honourable Mention in the Rolex Awards for Enterprise.

Whitaker realized long ago that snakes and the other species he loves cannot survive without their habitats. So, like many others, he has evolved from naturalist to conservationist. 'A lot of us get wrapped up in our own little special animal and then we wake up and start thinking it has got to be habitat and it has to be eco-development that involves people and, now, in my case, it has crystallized into the whole idea of water resources.

'India has a history of droughts, floods and famines,' Whitaker explains. 'Food production has been successfully tackled and dealt with, but we are now faced with a water shortage that will dwarf any of the past problems faced by the people. Owing to forest clearance and ill-advised dam projects, rivers are drying up, ground water reserves are being used up faster than they

The mountainous Western Ghats, once cloaked in lush rainforest, have been largely cleared for plantations. Only a few small patches of high rainfall areas feed the rivers that provide water for millions on the subcontinent. Water resources are a focus for Whitaker's research stations along the Ghats, in Assam and the Andaman and Nicobar islands. At Agumbe, where he caught his first king cobra, Whitaker bags and releases one he found near a house.

'Romulus Whitaker is without doubt one of the leading lights both in Indian herpetology and the Indian conservation movement today.'

DR WOLFGANG WÜSTER
SCHOOL OF BIOLOGICAL SCIENCES, UNIVERSITY OF WALES

can be replenished and pollution is hitting most of our sources of drinking water. These are the obvious problems, but there are other, possibly much more serious threats facing our water regimes including climate change, which we must tackle on a war footing.'

Ironically the water that Whitaker is intent on saving is – in the form of rain – one of the major obstacles to conservation research in many parts of India. Despite being recognized worldwide as biodiversity hotspots, relatively little is known about India's dwindling rainforests and the many species to which they are home. But monsoon downpours make it near impossible for researchers to operate at the very time the most scientifically interesting events are occurring in the landscape and the lives of its inhabitants. At Agumbe, where Whitaker caught his first king cobra back in 1971, annual rainfall of 10 metres or so condemns outsiders not just to swarms of leeches, along with wet clothes and tents, but also to guaranteed malfunction in all the equipment bound up in recording, communicating and calculating.

Whitaker's base at Agumbe, constructed in 2005, and now a fully functioning research, conservation and education centre, is the first of seven research stations that will connect key remaining rainforest strongholds throughout India. Sita Nadi, a river that has its source near the Agumbe Station, is a major focus for Whitaker and his team, who have started a small but ambitious plan to clean up and maintain the integrity of the river, using a three-pronged approach: evaluating the problems, involving the people and implementing a practical action plan. Whitaker cannot emphasize enough the importance of the region's rainforests for water resources. 'The rainforests of India are the origin of all the major rivers in the south and the north-east,' he points out. 'The rivers in the Western Ghats [in India's south] provide the water for 350 to 400 million people, about a third of India's population.'

The Agumbe station itself consists of living and working quarters purpose-built to function during the monsoon and to be self-sufficient in renewable energy. It is strategically located on about 3 hectares of land adjacent to a wildlife sanctuary and a national park so that field scientists have easy access to the forest. The base has hosted dozens of Indian researchers, journalists and naturalists. But the station's mission extends beyond science. It is a springboard for local conservation, including the sustainable use of minor forest produce and medicinal plants. The station has educated hundreds of schoolchildren about the forest. 'Children are a bit shaky about going into the forest at first, but fascination with what we show them soon gets them hooked,' Whitaker says.

The network of seven stations will produce vital information, building on discoveries by Whitaker's colleagues of over 100 new species of frogs in the last decade, and the study of crabs that live in trees. The network will allow immediate exchange of expertise and research, creation of a

'Research stations are the only means of promoting long-term research in the tropics, and studies allow us to document the biology and ecology of local organisms.'

GEORGE R. ZUG
SENIOR CURATOR AND RESEARCH ZOOLOGIST, SMITHSONIAN NATIONAL MUSEUM OF NATURAL HISTORY

comprehensive biodiversity database and expanded mobile educational programmes. Five of the stations in the network, including Agumbe, will be located in the states that span the Western Ghats. A sixth station will be located in the far north-eastern state of Assam, a vital haven for large numbers of migratory birds and endangered mammals. The seventh station is in the Andaman and Nicobar Islands, 350 tropical islands situated 1,200 kilometres from the Indian mainland in the Bay of Bengal.

Six of the seven stations already exist in various stages of development and now need vital new laboratory equipment and in some cases physical expansion to bring them up to speed for the network. Whitaker will use the Rolex Award to help make this happen. Only one station, near the Kalakkad-Mundanthurai Tiger Reserve in the southern tip of India, needs to be built from the ground up. The Rolex funds will also be used to fit out the bases with automatic weather stations. 'Climate change is tightly linked with the future of water resources and we need to be monitoring it now,' he says.

All those who know Whitaker's work agree that his ability to implement environmental projects is considerable. 'Rom Whitaker is passionate about conservation and he is an intrepid fieldworker,' says S. Theodore Baskaran, honorary wildlife warden and former Postmaster General of Tamil Nadu. 'As an institution builder, Rom is unfazed by any hurdle he might face in his work.'

Whitaker puts his trust not just in his own skills, but also in the aspirations of younger generations: 'We are doing a lot of work with young people, bringing them to the forest and showing them what happens here and why it matters. It can be very difficult to change adult attitudes, but with the young, it is easier to get across the knowledge that what we are doing to the forests we are doing to ourselves.' |

Research and education are the touchstones of Whitaker's work. From his Agumbe headquarters, top right, staff and volunteers go into the rainforest to gather information on a termite mound for a biodiversity database. Local school girls come to learn about snakes. In a land where they are often venerated, snakes still strike fear in many. Whitaker wants to teach them how valuable reptiles are to the ecosystem.

Phil Dickie WRITER   Cedric Bregnard PHOTOGRAPHER

Laureates and Associate Laureates 1978–2008

CATHERINE ABADIE-REYNAL	France	Excavate two ancient cities on Turkey's Euphrates before the area is flooded by an artificial lake
MOHAMMED BAH ABBA	Nigeria	Manufacture and supply an innovative earthenware food cooling system in poor, arid countries
NANCY ABEIDERRAHMANE	Mauritania	Develop the world's first industrial production of camel cheese in Mauritania
TALAL AKASHEH	Jordan	Conserve ancient Petra from the ravages of time and tourism
ERNA ALANT	South Africa	Expand a programme of non-oral communication to the disabled in South Africa's poor, rural areas
MICHEL ANDRÉ	France	Develop a whale anti-collision system to protect whales from collisions with ships
JOHN FRIEDRICH ASMUS	United States	Develop laser techniques to clean and restore colour to China's ancient terracotta army
JACQUES LUC AUTRAN	France	Bring medical and technical aid by ship to isolated communities on the islands of the Indian Ocean
JOSÉ MÁRCIO AYRES	Brazil	Protect the Amazon forest in Brazil by including local people in conservation
TIM BAUER	United States	Reduce pollution from motorized tricycles in Asian cities
ALEXIS BELONIO	Philippines	Turn rice husks into cheap, clean energy for cooking
CRISTINA BUBBA ZAMORA	Bolivia	Recover stolen Bolivian ceremonial weavings and return them to their original Andean communities
PISIT CHARNSNOH	Thailand	Prevent the endangered dugong from vanishing from Thai waters by preserving its habitat
IRINA CHEBAKOVA	Russia	Hold annual marches to boost public support to save Russia's protected natural areas
SEBASTIAN CHUWA	Tanzania	Lead a massive reforestation project to preserve the natural environment in northern Tanzania
GILBERT A. CLARK	United States	Enable students worldwide to access a network of professional telescopes and operate them remotely
SABINE COTTE	France	Produce a manual of simple conservation techniques for the monks at Bhutan's fortified monasteries
LAURY CULLEN JR.	Brazil	Transform farmers into conservationists to save the Atlantic forest and its fauna in eastern Brazil
SHEKAR DATTATRI	India	Save wild India through films by raising public and policy-makers' awareness on environmental issues
ANTONIO DE VIVO	Italy	Explore the underground rivers and caves of the Rio La Venta Canyon in southern Mexico
LUC DEBECKER	Belgium	Explore prehistoric lifestyles and beliefs through the study of European cave paintings
TOMAS DIAGNE	Senegal	Save Senegal's endangered land tortoises, Africa's largest, through a breeding sanctuary
SANOUSSI DIAKITÉ	Senegal	Invent a fonio-husking machine to revive cultivation of this healthy, inexpensive cereal in Africa
CRISTIAN DONOSO	Chile	Explore western Patagonia in sea kayaks to study and raise awareness of this little-known area
LONNIE DUPRE	United States	Undertake the first summer crossing of the Arctic Ocean to raise awareness of global warming
LUC-HENRI FAGE	France	Explore ancient cave paintings in Kalimantan Borneo to save these artistic treasures
CLAUDIA FEH	Switzerland	Establish an interactive learning forum to support the reintroduction of Przewalski horses in Mongolia
MARTINE FETTWEIS-VIÉNOT	Belgium	Create the first complete catalogue of Mayan wall paintings to study their role in Mayan society
ANABEL FORD	United States	Establish the El Pilar archaeological site as a model for conservation and sustainable development
BERNARD FRANCOU	France	Study El Niño and the impact of global warming through a climatic record preserved in Andean ice

STEVEN LURIE GARRETT	United States	Develop a sound-powered, CFC-free refrigerator to help reduce depletion of the ozone layer
ERIC GILLI	France	Study traces of ancient earthquakes in caves to develop a new methodology for predicting earthquakes
WIJAYA GODAKUMBURA	Sri Lanka	Replace makeshift kerosene-oil bottle lamps with safe lamps to prevent burns
ZENÓN GOMEL APAZA	Peru	Revive traditional Andean agriculture to ensure food security and strengthen ties between communities
ARTURO GONZÁLEZ	Mexico	Explore submerged caves to discover and study remains from the Ice Age
GORUR R. IYENGAR GOPINATH	India	Expand ecological silk-farming in India to preserve the environment and improve living standards
RAFAEL GUARGA	Uruguay	Protect crops from frost through use of an innovative, inexpensive cold-extracting chimney
ROYCE O. HALL	United States	Build an eye hospital in Tanzania and bring state-of-the-art surgery procedures to the African bush
KENNETH WILLIAM HANKINSON	United Kingdom	Undertake the first over-winter stay in tents in Antarctica's Brabant Island and make pioneering studies
HANS HENDRIKSE	South Africa	Produce a low-cost rolling water-drum to allow efficient water transfer in developing countries
GEORGINA HERRMANN	United Kingdom	Explore the Silk Road caravan city of Merv in Turkmenistan
SHAFQAT HUSSAIN	Pakistan	Protect snow leopards with an innovative scheme combining livestock insurance and ecotourism
DAVE IRVINE-HALLIDAY	Canada	Supply inexpensive, reliable, low-powered LED lighting systems in developing countries
RODNEY M. JACKSON	United Kingdom	Protect the endangered Himalayan snow leopard through radio-tracking and study
NORBERTO LUIS JÁCOME	Argentina	Pioneer efforts to protect the Andean condor
RUNA KHAN MARRE	Bangladesh	Preserve the ancient boat-building heritage of Bangladesh by opening a 'living museum'
PETER KNIGHTS	United Kingdom	End the consumption of products from endangered species through public awareness campaigns
ILSE KÖHLER-ROLLEFSON	Germany	Use traditional and modern knowledge to save camels in Rajasthan while protecting the Raika way of life
KAREL KOLOMAZNIK	Czech Republic	Improve technology to recover and recycle potentially toxic wastes from the leather industry
STEPHEN W. KRESS	United States	Re-establish endangered seabird colonies through innovative 'social attraction' techniques
BILLY LEE LASLEY	United States	Facilitate the breeding of endangered birds by developing a non-invasive method to identify their sex
ALEXANDRA LAVRILLIER	France	Run a nomadic school in Siberia to provide Evenk children an education blending tradition and modernity
NABIL M. LAWANDY	United States	Develop an inexpensive photodynamic treatment for certain forms of cancer
LOUIS LIEBENBERG	South Africa	Develop a hand-held computer that modernizes ancient tracking skills and improves wildlife management
EDUARDO LLERENAS	Mexico	Trace and record Mexico's traditional folk music to preserve the country's rich musical heritage
ALDO LO CURTO	Italy	Produce a practical, illustrated health education manual for the Indians of Brazilian Amazonia
DAVID LORDKIPANIDZE	Georgia	Explore and protect the earliest known site of human activity outside Africa, in Dmanisi, Georgia

TERESA MANERA	Argentina	Preserve prehistoric animal tracks at a unique palaeontological site along the Atlantic coast of Argentina
MARIA ELIZA MANTECA OÑATE	Ecuador	Promote sustainable farming in the Ecuadorian Andes to improve the environment and living standards
KENNETH LEE MARTEN	United States	Save the Abyssinian wolf in Ethiopia by developing a conservation programme
ANDREW McGONIGLE	United Kingdom	Predict volcano eruptions using a remote-controlled helicopter
RODRIGO MEDELLÍN	Mexico	Save endemic and endangered bats through habitat protection and education
JULIEN MEYER	France	Revitalize and preserve whistled and drummed languages via the Internet
FORREST MARION MIMS III	United States	Set up a global network to monitor ultraviolet radiation and ozone levels with an innovative ozonometer
KIKUO MORIMOTO	Japan	Revive traditional silk production in Cambodia to provide a sustainable income for rural communities
PIERRE MORVAN	France	Study ground beetles in Nepal to deepen understanding of how biological species are formed
MAKOTO MURASE	Japan	Recycle rainfall to solve urban water shortages and provide a safe, sustainable supply of water
ANDREW MUIR	South Africa	Provide training and jobs to young people orphaned by Aids
NANCY LEE NASH	United States	Broaden the understanding of conservation by researching and applying Buddhist precepts on nature
ELIZABETH NICHOLLS	Canada	Extract the fossilized remains of a 220-million-year-old giant marine reptile in remote Canada
DORA NIPP	Canada	Preserve a collection of oral testimonies by Canadian immigrants to enhance multiculturalism
BRAD NORMAN	Australia	Establish a global photo-identification network to help conserve and understand the whale shark
FRANCINE PATTERSON	United States	In a sanctuary in Hawaii, test the ability of gorillas to learn and use human sign language
JEAN-FRANÇOIS PERNETTE	France	Explore and map the earth's most southerly underground caves on the remote islands of Patagonia
DONALD RAY PERRY	United Kingdom	Study the rainforest canopy above the Costa Rican jungle floor through a system of ropes and cables
ROHAN PETHIYAGODA	Sri Lanka	Create rainforest microcosms in Sri Lanka as living laboratories to protect a threatened biodiversity
PILAI POONSWAD	Thailand	Engage rural communities in saving threatened hornbills and their rainforest habitat in Thailand
LAURENT PORDIÉ	France	Promote the ancient Amchi system of Tibetan medicine to protect traditional communities in Ladakh
SURYO PRAWIROATMODJO	Indonesia	Establish Indonesia's first environmental education centre and pioneer an innovative model for this field
ADLI QUDSI	Syria	Preserve and restore the Old City of Aleppo, Syria, and rebuild a thriving community
JOHAN REINHARD	United States	Preserve the patrimony of the Andean people through high-altitude cultural anthropology and archaeology
MOJI RIBA	India	Safeguard the heritage of the people of Arunachal Pradesh
MARIO ROBLES DEL MORAL	Spain	Establish a national reforestation programme in Spain in light of a major soil erosion problem

LINDY RODWELL	South Africa	Preserve African cranes and their wetland habitats by building a network of crane conservationists
WILLIAM ROSENBLATT	United States	Channel unused surgical supplies from hospitals in affluent countries to those in developing countries
MARTHA RUIZ CORZO	Mexico	Pioneer efforts to combine conservation with economic development in Mexico's Sierra Gorda Mountains
GORDON SATO	United States	Help Eritreans to establish a multifaceted agricultural programme through mangrove cultivation
VALERIO SBORDONI	Italy	Explore the remote underground caves of southern Mexico for biodiversity and new life forms
DAVID SCHWEIDENBACK	United States	Redistribute used bicycles to stimulate economic growth and increase mobility in developing countries
JUNICHI SHINOZAKI	Japan	Research the environment in the mountains of the Pacific Rim to determine pollution levels
CHANDA SHROFF	India	Revive the rich, ancient tradition of embroidery to create a sustainable income for rural women in Kutch
ALEXANDER STANNUS	United Kingdom	Circle the earth via inland waterways and allow schoolchildren to follow the expedition interactively
LES STOCKER	United Kingdom	Create Europe's first-ever wildlife teaching hospital in England to care for injured animals
ANITA STUDER	Switzerland	Develop a long-term reforestation and environmental education programme in north-east Brazil
GEOFFREY SUMMERS	United Kingdom	Apply innovative archaeological methods to map and explore Turkey's Iron-Age city of Kerkenes
THEAN SOO B. TEE	Malaysia	Grow asparagus on Mt Kinabalu to prevent soil erosion and create a new cash crop in the state of Sabah
MICHEL TERRASSE	France	Reintroduce griffon vultures to their natural habitat in the Massif Central region of southern France
JO THOMPSON	United States	Rebuild a war-ravaged field base in the Congo to continue community-based conservation of the bonobo
GEORGE VAN DRIEM	Netherlands	Chart and study the indigenous languages of the Himalayan region to help conserve them
AMANDA VINCENT	Canada	Protect seahorses in the Philippines and develop an alternative livelihood for fishing communities
FRITHJOF VOSS	Germany	Use satellites to control crop-devouring locust swarms in Africa before they reach a critical mass
ROMULUS WHITAKER	India	Establish a network of rainforest research stations across India
RORY WILSON	United Kingdom	Develop a revolutionary electronic logging device to track animals in the wild and analyse their behaviour
REUVEN YOSEF	Israel	Create a sanctuary in Eilat, Israel, as a stopover for birds migrating between Eurasia and Africa
ELSA ZALDÍVAR	Paraguay	Combine loofah and plastic waste to make low-cost housing

The 2008 Rolex Awards Selection Committee

Eminent specialists who themselves embody the spirit of enterprise

'Being on the Selection Committee allows you to extend opportunities to those passionate, curious people who are … pushing back the barriers of knowledge.'

DR KATHRYN D. SULLIVAN
FORMER NASA ASTRONAUT AND DIRECTOR OF THE BATTELLE CENTER FOR MATH AND SCIENCE EDUCATION POLICY

The winners of the Rolex Awards for Enterprise are chosen by the Selection Committee, a group of independent, multidisciplinary specialists who volunteer their services to select, from among the best entries, the ten projects that have the most resonance in meeting the stringent criteria: they consider whether the projects are feasible and original, and positively impact on the surrounding community and the world at large. Above all, they look for a demonstrable spirit of enterprise by the candidate as they scrutinize and debate the merits of each application: has he or she overcome the most challenging odds with ingenuity and bold determination?

In the 2008 series, 1,477 entries were received from 127 countries. First analysed by a team of scientific researchers at Rolex headquarters in Geneva in a rigorous, 10-month evaluation process that entailed checking facts and seeking expert opinions from knowledgeable consultants worldwide, the projects were then submitted to the Selection Committee.

Although the judging panel normally changes for every series, to mark the 30-year anniversary of the first prize-giving ceremony in 1978, six former Selection Committee members were asked to serve again.

The 2008 jury met in Geneva in April this year under the chairmanship of Rolex Chief Executive Officer Patrick Heiniger. These dozen men and women, like the nearly 100 judges who have served in past series, represent a broad spectrum of professions and nationalities. As entrepreneurs and educators, doctors and environmentalists, geologists and explorers, oceanologists, economists and astronauts, they are eminently qualified to choose the five Laureates and five Associate Laureates. Throughout their careers, all of these internationally renowned figures have themselves been driven by a pioneering spirit, often carrying out groundbreaking work under difficult conditions, and have today attained leadership positions in the general award areas of science, technology, the environment, preservation and exploration. They have also gained recognition for their wider interests and for their contributions to human knowledge and endeavour.

The five 2008 Laureates – those who demonstrated the most exceptional spirit of enterprise – were singled out by the judging panel and each granted US$100,000 and a gold Rolex chronometer. They were honoured at an official awards ceremony in Dubai, held under the patronage of HRH Princess Haya Bint Al Hussein, who is married to HH Sheikh Mohammed Bin Rashid Al Maktoum, Vice-President and Prime Minister of the UAE and Ruler of Dubai. The five Associate Laureates, the runners-up, each receive $50,000 and a steel-and-gold Rolex chronometer at ceremonies in their home countries or regions.

The Rolex Awards are notable for the diversity of the individuals they support and for the fact that, unlike other award schemes, prizes are given for ongoing, concrete projects rather than for past achievements. Anyone of any age, nationality or background with an original project to help mankind can apply. While winners welcome the funding, which must be used to implement or complete their innovative projects, they often cite the global recognition and publicity they receive as the greatest benefit. This acknowledgement boosts their international standing and accreditation, and often attracts additional funding, allowing them to expand their projects beyond all expectations.

Cedric Bregnard PHOTOGRAPHER

VIKRAM AKULA

India and the United States

Microfinance expert and founder and CEO of SKS Microfinance

Vikram Akula has gained respect worldwide for his innovative approaches to helping poor people improve their lives. Born in India, Akula grew up in the United States where his family emigrated in the 1970s. As a child, he often visited India, and the poverty that plagues the country left a lasting impression.

In 1995, thanks to a Fulbright Scholarship, he coordinated a microcredit programme in India. While doing this, he witnessed the tremendous impact that microfinance made on the lives of the poor. But he also realized that microfinance was not reaching its full potential because it was not able to scale to large numbers.

This inspired him to set up SKS Microfinance in Hyderabad in 1998 to economically empower the poor by providing them finance to establish micro-enterprises – and to do so at a scale that had not been done before. By applying principles of commercialization, operational excellence and technological innovation to microfinance, SKS has been able to provide close to US$600 million in micro-loans and micro-insurance services to over 2 million poor women across India – and outreach is growing at a rate of 200 per cent a year.

ETIENNE BOURGOIS

France

Head of agnès b. fashion company and Arctic expedition leader

An experienced expedition-leader and inveterate yachtsman, Etienne Bourgois has been head of the agnès b. fashion company for over 20 years. Born into a seafaring family, he was introduced to yachting by his grandfather and his uncle, Bruno Troublé, and went on to combine his passion for sailing with a commitment to protecting the environment.

Since 2004, Bourgois has led six expeditions in the polar schooner *Tara* to Greenland and the Arctic, enabling several teams of scientists, as well as artists, to raise awareness of the risks to the environment. The *Tara* is supported by the United Nations Environment Programme.

From September 2006 to January 2008, he supervised the Tara-Arctic Expedition. Working with the scientific European Union programme DAMOCLES (Developing Arctic Modelling and Observing Capabilities for Long-term Environmental Studies), the expedition's goal was to draw up a full inventory of the state of this fragile habitat and to gain a better understanding of the impact of global warming.

For Etienne Bourgois, this scientific – and personal – adventure is 'a commitment to the planet. It's my responsibility – as the head of a clothing company, and also as a citizen of a rich country – to increase people's awareness of the risks facing future generations.'

DENISE BRADLEY

Australia

President of the Australian College of Educators and campaigner for better educational standards

Denise Bradley, former vice chancellor and president of the University of South Australia (UniSA) and current president of the Australian College of Educators, is renowned for her unflagging efforts to raise the standard of higher education in Australia and abroad.

Bradley began work as a secondary school teacher in the early 1960s and soon became disillusioned with the barriers faced by women teachers. In the mid-1970s, she set out to improve women's education and employment opportunities and, by the end of the decade, she had been appointed a government advisor on women's education. In 1997, she became head of UniSA, the third woman ever selected as a president of an Australian university. She retired in May 2007, but, in March 2008, the newly elected federal government appointed her to chair a panel to develop a blueprint for higher education in Australia until 2020.

She has served on national and international bodies, advising governments on educational policy. 'We must endeavour to make education truly open to all,' says Professor Bradley. She is chair of the board of IDP Education Australia, a not-for-profit organization and world leader in international education.

Professor Bradley's ability to bring about change has won wide recognition. 'I wanted to change things and I was born at the right time for that to be possible for a woman,' she says.

GEH MIN

Singapore

Ophthalmologist, environmentalist and president
of the Nature Society of Singapore

A distinguished eye surgeon, Dr Geh Min is also one of South-East Asia's most committed conservationists. The first woman president of the Nature Society of Singapore (since 2000) and a nominated member of Parliament from 2004 to 2006, she also serves on the boards of the Singapore Environment Council and of the Nature Conservancy's Asia Pacific Council.

Geh Min is a tireless campaigner who has helped implement milestone environmental preservation policies such as the Singapore Green Plan 2012, a ten-year blueprint to build a sustainable environment. Under her leadership, the Nature Society of Singapore has been instrumental in campaigning for and helping to establish wetland and marine reserves.

Geh Min's passion for nature can be traced to her childhood, when her father took her walking in the jungle and deep-sea fishing. In 1980, she became a fellow of the Royal College of Surgeons in Edinburgh. Since 1988, she has practised as a consultant eye surgeon at Singapore's Mount Elizabeth Medical Centre at the National University Hospital and at the Singapore National Eye Centre.

She is also committed to promoting women's and children's rights, culture and the arts, and is an enthusiastic committee member of the Singapore chapter of the International Women's Forum. 'Whether it's creativity in the sciences or the arts, you finally have to get back to nature as the bottom line,' she says.

FARKHONDA HASSAN

Egypt

Geologist, professor of geology, and member
of second house of the Egyptian Parliament

A geologist, politician and development specialist, Farkhonda Hassan is widely respected for her unwavering commitment to science, nature conservation and to the advancement of women's rights throughout the Arab world, particularly as a member of the Executive Board of the Arab Women Organization.

She credits her father with what she calls her 'fascination' with geology – by showing her a fossil and explaining how it was formed, he sparked off a vocation that led to her becoming one of the few women studying geology in Egypt. After raising her two children, she resumed her studies, obtaining a Ph.D. in 1970 at the University of Pittsburgh, in the United States.

Alongside her research and teaching as geology professor at the American University in Cairo, Dr Hassan has served in national and international political institutions. Elected to the Egyptian Parliament in 1979, she played a key role in setting up the Egyptian Environmental Affairs Agency and in the implementation of environmental protection laws.

Professor Hassan has promoted women's rights and women's role in science as chair of the National Committee for Women in Science, president of the Scientific Association for Egyptian Women and co-founder of the Arab Network for Women in Science and Technology.

She has been co-chair since 1998 of the Gender Advisory Board of the United Nations Commission on Science and Technology for Development.

RODRIGO JORDÁN

Chile

Educator, mountaineer and Chairman of Chilean
National Foundation for the Alleviation of Poverty

Dr Rodrigo Jordán, leader of the first successful South American expedition to Mt Everest and K2, has applied the leadership and team-building skills needed to climb the world's most challenging mountains to business and education.

His mountaineering successes gave him the idea of launching in 1992 his own company, Vertical. This organization, along with the charitable foundation, Fundación Vertical, delivers outdoor education and training services to corporations and individuals, particularly children from inner-city areas.

By involving children in nature conservation, Jordán hopes to make future generations more aware of the environment. 'Those who have not had the opportunity of experiencing and enjoying nature as children will not understand the ethical and economic reasons for protecting it as adults,' he explains. A civil and industrial engineer, Jordán earned a Ph.D. in Organisational Development from Oxford University and is today a lecturer in Leadership at the Pontifical Catholic University of Chile.

His latest endeavours in exploration include an unsupported 400-kilometre crossing of the Ellsworth Mountains in Antarctica and an ascent of Lhotse, the world's fourth-highest mountain. In 2008, he took part in an international kayaking and mountaineering expedition to document climate change.

YOLANDA KAKABADSE

Ecuador

Environmentalist and global champion of
sustainable development

'My heart is in nature conservation,' says Yolanda Kakabadse, one of today's most prominent environmentalists, widely recognized as a global champion of sustainable development and bio-diversity preservation.

Yolanda Kakabadse was born in Quito, Ecuador, where she studied educational psychology before turning to nature conservation in 1979 as executive director of Fundación Natura. Under her leadership, the foundation gained great influence, contributing largely to policies promoting the sustainable use of natural resources.

Kakabadse then coordinated the participation of non-governmental organizations at the 1992 Earth Summit in Rio. The following year, she established the Fundación Futuro Latino-americano, of which she was president until 2006. She is now chair of its Advisory Board. The foundation brings together environmental, political, cultural, industrial and economic sectors to prevent social-environmental conflicts in Latin America.

From 1996 to 2004, she presided over the World Conservation Union (IUCN). She was Ecuador's Minister of Environment from 1998 to 2000, and co-chaired the Environmental Sustainability Task Force of the United Nations Millennium Project from 2002 to 2005. Kakabadse now chairs the scientific and technical advisory panel (STAP) of the Global Environment Facility (GEF).

R.T. (PHIL) NUYTTEN

Canada

Inventor, entrepreneur, deep-ocean explorer and
President and founder of Nuytco Research Ltd

An internationally recognized pioneer in the undersea industry, Phil Nuytten has spent 40 years creating deepwater dive systems that have opened the ocean's depths to exploration and industry.

Through his various companies, he has developed technology to allow deeper and longer-length underwater expeditions with increased safety. Nuytten's hard-suits – the Newtsuit and the Exosuit – his deep-diving submersibles and his submarine rescue system are world renowned. His equipment is used by a wide range of agencies from the National Geographic Society to NASA and is standard in nearly a dozen navies. Contract assignments have taken Nuytten's equipment and crews to oilfields, marine construction sites and sunken wrecks worldwide.

Born in Vancouver, British Columbia, Nuytten designed his first diving gear and opened Western Canada's first dive shop when he was in his teens. His attention to detail – crucial in deep-sea diving – was apparent from the age of 12 when he apprenticed with a Kwakiutl master totem-pole carver and learned to carve and design in the Northwest Coast style – an art about which he is still passionate. Nuytten is a Métis (an aboriginal group descended from Canadian fur traders) by birth, but was formally adopted into the Kwakiutl nation and is an advocate of the art and culture of these indigenous people.

IVO PITANGUY

Brazil

Plastic surgeon, professor and director of the
Ivo Pitanguy Clinic and Institute

Recognized as one of the world's most eminent plastic surgeons, Ivo Pitanguy has enhanced the appearance of thousands of people, including hundreds of the world's celebrities. He has also gained deep respect for performing reconstructive surgery on the poor, and for his dedication to creating a better life for those who are burned or maimed or have birth defects.

Educated in Brazil, the U.S. and France, he was a pioneer of plastic surgery in Latin America and has been practising in Rio de Janeiro since the 1950s.

'For a patient, even a small physical deformity can create a great human or social loss. My role is to help those who come to me by giving them the hope of having a normal life,' he says.

Dr Pitanguy has published more than 900 scientific papers about plastic surgery and related issues. Eager to share his knowledge, he is the chief professor of the plastic surgery departments of the Carlos Chagas Medical School and the Catholic University of Rio de Janeiro. He has taught since 1960, training over 500 plastic surgeons from 40 countries.

In 1964, he established a plastic surgery study centre that later became the Pitanguy Institute, where more than 60,000 patients have been treated.

An ardent nature-lover, Dr Pitanguy created an ecological sanctuary on the island of Porcos Grande in the 1970s.

ANATOLY M. SAGALEVITCH

Russia

Oceanologist, head of the Laboratory of Manned
Submersibles, P.P. Shirshov Institute of Oceanology

A pioneering explorer of deep oceans, Dr
Anatoly Sagalevitch has spent thousands of
hours underwater. As head of the manned
submersibles laboratory at Moscow's P.P.
Shirshov Institute of Oceanology, where he has
worked for 42 years, Dr Sagalevitch helped
design the twin MIR submersibles that have
carried him and his international team of scientists to depths of more than 6,000 metres.

Since 1990, the eminent oceanologist has
served as chief scientist and expedition leader
on board the MIR support vessel, piloting the
submersibles on dives worldwide. These underwater operations include expeditions to study
hydrothermal vents, fissures in the ocean floor;
to investigate the nuclear submarines *Komso-
molets* and *Kursk* after accidents stranded them
on the seabed; and to film the wreck of the
Titanic. The numerous scientific missions led by
Dr Sagalevitch have broadened knowledge of
the oceans and paved the way for further collaboration between Russian and foreign scientists.

In 2007, he made the first-ever descent to
the bottom of the North Pole, at a depth of 4,300
metres.

'A unique spirit of cooperation exists among
the world's oceanographers, especially among
the few of us who journey to the ocean depths,'
says Sagalevitch. '*Mir*, meaning "peace" in Russian, is an apt name for the submersibles.'

EMIL SALIM

Indonesia

Professor of economics at the University
of Indonesia

Dr Emil Salim is a leading voice on environmental issues in his native Indonesia and abroad.
A distinguished economist, he has taught economics at the University of Indonesia since 1972.
He was minister of the environment from 1978
to 1993, and enacted his country's first comprehensive law on the environment and sustainable
development in 1982.

From 1995 to 1999, he co-chaired the United
Nations High-Level Advisory Board for Sustainable Development. He was also the founder
and chairman of the Indonesian Biodiversity
Foundation and of the country's Ecolabeling
Institute, a group which ensures the ecological
soundness of forestry products.

He has been a member of high-level boards
on population, health and conservation for the
United Nations, the World Commission on Forests
and Sustainable Development, the World Bank
and the World Health Organization, and an Eminent Person on the Extractive Industries Review.

Dr Salim's devotion to conservation can be
traced to his days as a young man living on the
densely forested island of Sumatra where he
became aware of Indonesia's bounty of rich,
natural resources – and the threat posed by
indiscriminate development.

'It is each generation's responsibility to
maintain the environment for those who follow,'
he says. 'This is often ignored in the developing
world where natural resources are abused. We
must work to reverse this situation.'

KATHRYN D. SULLIVAN

United States

Director of the Battelle Center for Math and
Science Education Policy at Ohio State University

An illustrious scientist and explorer of sea and
space, Dr Kathryn Sullivan has made history with
her pioneering journeys to the world's frontiers.

'I spent my childhood in California exploring the wide, open spaces near my home and
feeding my scientific instincts,' says Sullivan.
Further inspired by her aerospace engineer
father and educated in earth sciences, she was
one of six women to join the first class of space-
shuttle astronauts in 1978. She is a veteran of
three missions: during a *Challenger* flight in
1984, she became the first American woman to
walk in space; she helped launch the Hubble
Space Telescope aboard *Discovery* in 1990; and
she served as a payload commander aboard
Atlantis in 1992.

After 15 years with NASA, she was appointed
chief scientist at the National Oceanic and
Atmospheric Administration, overseeing research
on issues from climate change to marine
biodiversity.

In 1996, Sullivan became president and
CEO of Ohio's Center of Science and Industry
(COSI), a learning centre for science, math and
technology. Over the next decade, she led COSI
through the most significant growth in its history.

She is now inaugural director of the Battelle
Center for Math and Science Education Policy at
Ohio State University, set up to increase the number
of students with strong backgrounds in science,
technology, engineering and mathematics.

Selection Committee Members 1978–2008

PATRICK HEINIGER	Switzerland	*Chairman*, 2000–
ANDRÉ HEINIGER	Switzerland	*Chairman*, 1976–1998
LARETNA T. ADISHAKTI	Indonesia	Architect
VIKRAM AKULA	India/United States	Microfinance expert
MARY ARCHER	United Kingdom	Chemist
JUAN LUIS ARSUAGA	Spain	Palaeoanthropologist and biologist
ORIOL BOHIGAS GUARDIOLA	Spain	Architect
ROBERTA BONDAR	Canada	Neurologist and astronaut
ETIENNE BOURGOIS	France	Entrepreneur and expedition leader
DENISE BRADLEY	Australia	Educator
TEODOSIO CÉSAR BREA	Argentina	Lawyer
CHARLES F. BRUSH	United States	Anthropologist
MARGARET BURBIDGE	United Kingdom	Astrophysicist
ADRIENNE ELIZABETH CLARKE	Australia	Biochemist
GEORGE VAN BRUNT COCHRAN	United States	Orthopaedist and explorer
YVES COPPENS	France	Palaeoanthropologist and prehistorian
FLEUR COWLES	United Kingdom	Painter and author
WALTER CUNNINGHAM	United States	NASA astronaut
NILS DAHLBECK	Sweden	Ecologist
LAURENCE DE LA FERRIÈRE	France	Alpinist and explorer
JOËL DE ROSNAY	France	Scientist
SANTIAGO DEXEUS	Spain	Gynaecologist
JEAN DORST	France	Ornithologist
SYLVIA A. EARLE	United States	Marine botanist and aquanaut
LAILA EL-HAMAMSY	Egypt	Cultural anthropologist
LEO ESAKI	Japan	Physicist
MARTA ESTRADA	Spain	Oceanographer and marine ecologist
RENÉ G. FAVALORO	Argentina	Cardiovascular surgeon
XAVIER FRUCTUS	France	Specialist in hyperbaric physiology
KATHRYN FULLER	United States	Lawyer and nature advocate
WILLIAM GRAVES	United States	Editor
SUSAN GREENFIELD	United Kingdom	Neuroscientist and pharmacologist
GILBERT M. GROSVENOR	United States	Chairman of National Geographic Society Board of Trustees
FARKONDA HASSAN	Egypt	Geologist
EDMUND HILLARY	New Zealand	Mountaineer
HEISUKE HIRONAKA	Japan	Mathematician
JOHN HUNT	United Kingdom	Former president of Royal Geographical Society and member of the House of Lords
YOSHIMINE IKEDA	Brazil	Oceanographer
MOTOKO ISHII	Japan	Lighting designer
DEREK A. JACKSON	France	Nuclear physicist
GUILLERMO JAIM-ETCHEVERRY	Argentina	Neurobiologist and educator
RODRIGO JORDÁN	Chile	Educator and explorer
ERLING KAGGE	Norway	Polar explorer and publisher
YOLANDA KAKABADSE	Ecuador	Environmentalist
MOHAMED KASSAS	Egypt	Desert ecologist
FRANCISCO KERDEL VEGAS	Venezuela	Dermatologist
REINER KLINGHOLZ	Germany	Molecular biologist and science editor
PATRICIA KOECHLIN-SMYTHE	United Kingdom	Olympics equestrian